华域普风

深圳市华域普风设计有限公司（POF），由一批在海外学习建筑学后归国实践的建筑师于2008年在深圳创立，是一家注重创意及品质，并始终坚持将国际化的设计理念与本土实践相结合的设计机构。公司主创团队成员均有丰富的设计实践及项目管理经验，其下各团队在设计细分领域各有专长。

在城市化的大时代背景下，普风注重对项目环境进行独立思考，并努力将自身的理念转化为某种设计条件。在设计周期逐渐被压缩的现实中，我们坚持对项目多种可能性的努力探索以及对最终呈现的效果做出各种有效的应对和进行更多的建造指引。

历经十载耕耘，普风作品已分布深圳、成都、重庆、南京、长沙、合肥、南宁、贵阳、长春、海口、三亚、珠海、东莞、惠州等二十几个城市，并获得市场的高度认可及众多设计奖项。

目 录
Contents

海南中信国安七星海岸总体规划

海南 万宁 | 2017

Master Plan of CITIC Guo' an Seven-star Coast, Hainan

Wanning, Hainan | 2017

性质：综合建筑
用地面积：754 760 m²
建筑面积：654 490 m²
开发商：海南明远置业有限公司
主题词：化零为整；水域；文化旅游；运营规划

Property: Complex Architecture
Site Area: 754,760 m²
Floor Area: 654,490 m²
Client: Hainan Mingyuan Real Estate Co., Ltd.
Keywords: Gathering Parts into a Whole;
Water; Cultural Travel; Operating Plan

San Marco
San Marco
DKNY
DKNY
DKNY

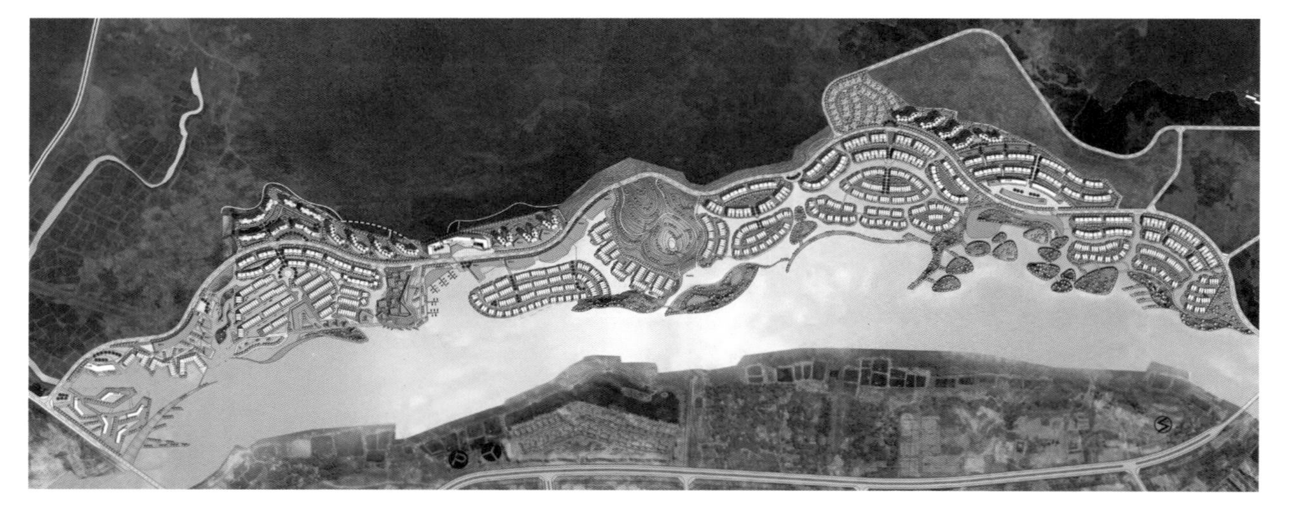

有别于传统的开发模式，该项目立足于新型文旅度假生活，坚持“山”“海”“城”统一规划的思路，着力打造海南岛内首个山海一体、依山面海的滨海度假区。规划合理利用了3.5 km长的海岸线，化零为整，以“共享海湾”为主题，构建系统化、分板块的多中心城市空间系统，并强调项目与周边城市要素的关联性。

在海岸线处理上，将绵长地块切割成尺度合理的居住小岛，创造生态宜居的水城生活。另外，项目空间规划与产业落地及运营相结合，并引入国安产业资源，为海湾生活及城区发展注入活力。

Different from the traditional development model, the project, based on the new style cultural travel holiday life，adheres to the idea of unifying the “mountain”，“sea” and “city”, and builds the first coastal resort close to the mountain and sea within Hainan Island. With the theme of sharing the bay, the plan gathers parts into a whole, and makes reasonable use of 3.5 km-coastline to create a systematic sub-plate multi-center urban space system, and emphasizes the relationship between the project and the surrounding urban elements.

The long block is cut into small islands for living with reasonable scale to create ecological livable water life. In addition, the integration of space planning with industry drop-off and project operation, and the introduction of the industry resources of Guo' an, provide vitality for the gulf life and urban development.

海南中信国安七星海岸规划与建筑设计

海南 万宁 | 2017

CITIC Guo’an Seven-star Coast Planning and Architectural Design, Hainan

Wanning, Hainan | 2017

性质：综合建筑
用地面积：200 937.00 m^2
建筑面积：166 298.67 m^2
开发商：海南明远置业有限公司
主题词：城市客厅；街区商业；高端海居

Property: Complex Architecture
Site Area: 200,937.00 m^2
Floor Area: 166,298.67 m^2
Client: Hainan Mingyuan Real Estate Co., Ltd.
Keywords: City Living Room；Block Business；
High-end Sea Residence

俏江南
眞 MANABE 鍋
東方御宴

地块理念与整体规划设计一脉相承，各个地块有机互联，共同塑造优质的度假生活，打造活力海岸。项目提出城市客厅的概念，为城市留出各类型的公共场地，创造共享空间，激发文化活力。因地制宜，设计港湾式街区商业，使商业活动灵活有机地融入到公共环境中。小区规划大胆创新，设计建筑间距宽裕的棋盘式布局，最大化利用场地优势，顺应地势和景观面，打造观山阅海的生活品质。

The concept of land parcel is traced to the same origin of overall planning and design. Each land parcel is organically connected with another to shape the quality of life and to create a vibrant coast. The project adopts the concept of urban living room, setting aside all types of public venues for the city, creating shared space and stimulating cultural vitality. Harbor-style block business, which is designed according to local conditions, makes commercial activities flexibly and organically integrated with the public environment. Bold and innovative community planning designs a well-spaced checkerboard layout with ample architectural space, and creates the life quality of viewing the mountain and sea by maximizing the use of site advantages and adapting to the topography and landscape surface.

Allstate
ForeSource
BAKERY
Albertsons
GURUZ

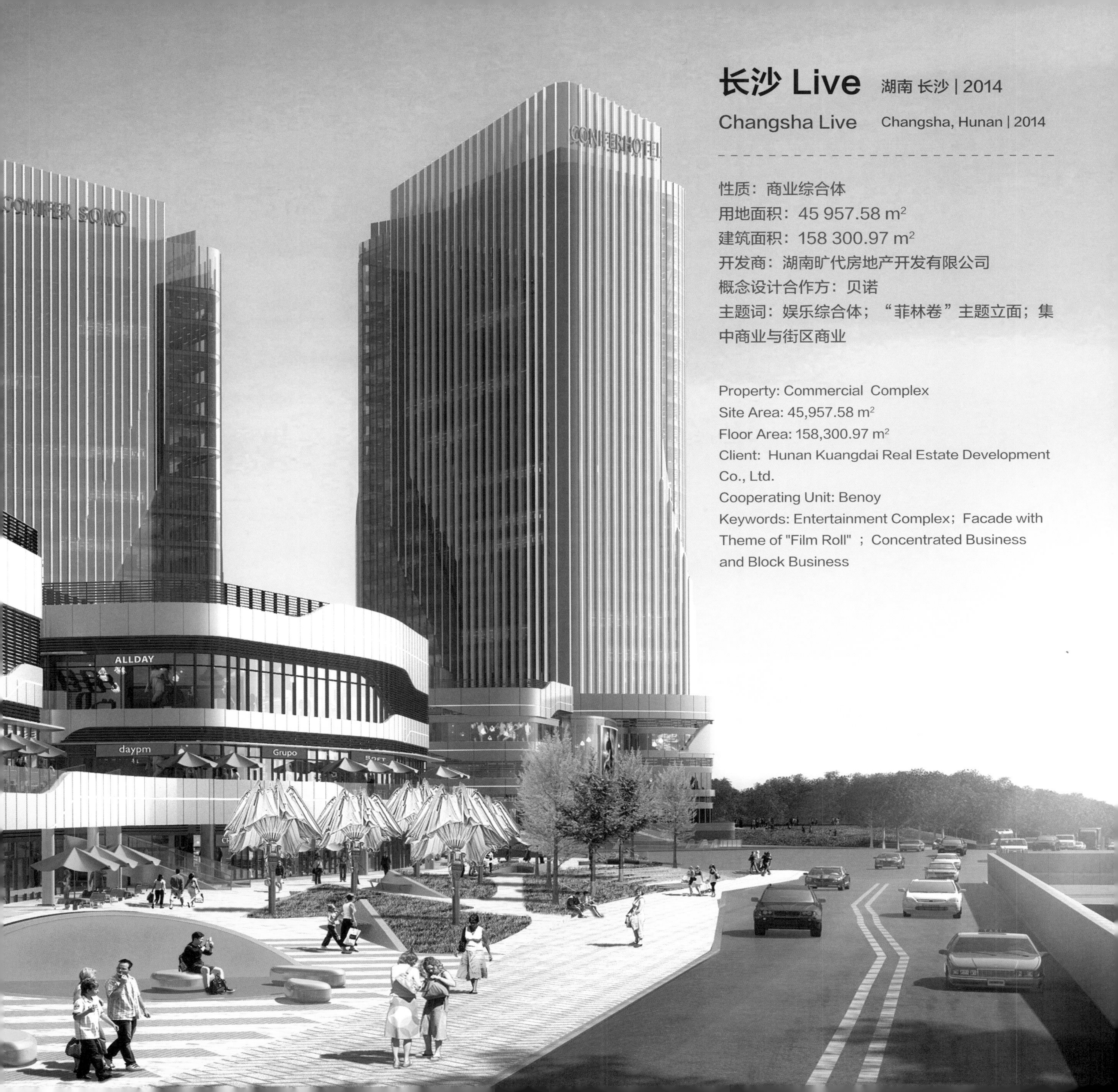

长沙 Live 湖南 长沙 | 2014

Changsha Live Changsha, Hunan | 2014

性质：商业综合体
用地面积：45 957.58 m^2
建筑面积：158 300.97 m^2
开发商：湖南旷代房地产开发有限公司
概念设计合作方：贝诺
主题词：娱乐综合体；“菲林卷”主题立面；集中商业与街区商业

Property: Commercial Complex
Site Area: 45,957.58 m^2
Floor Area: 158,300.97 m^2
Client: Hunan Kuangdai Real Estate Development Co., Ltd.
Cooperating Unit: Benoy
Keywords: Entertainment Complex；Facade with Theme of "Film Roll" ；Concentrated Business and Block Business

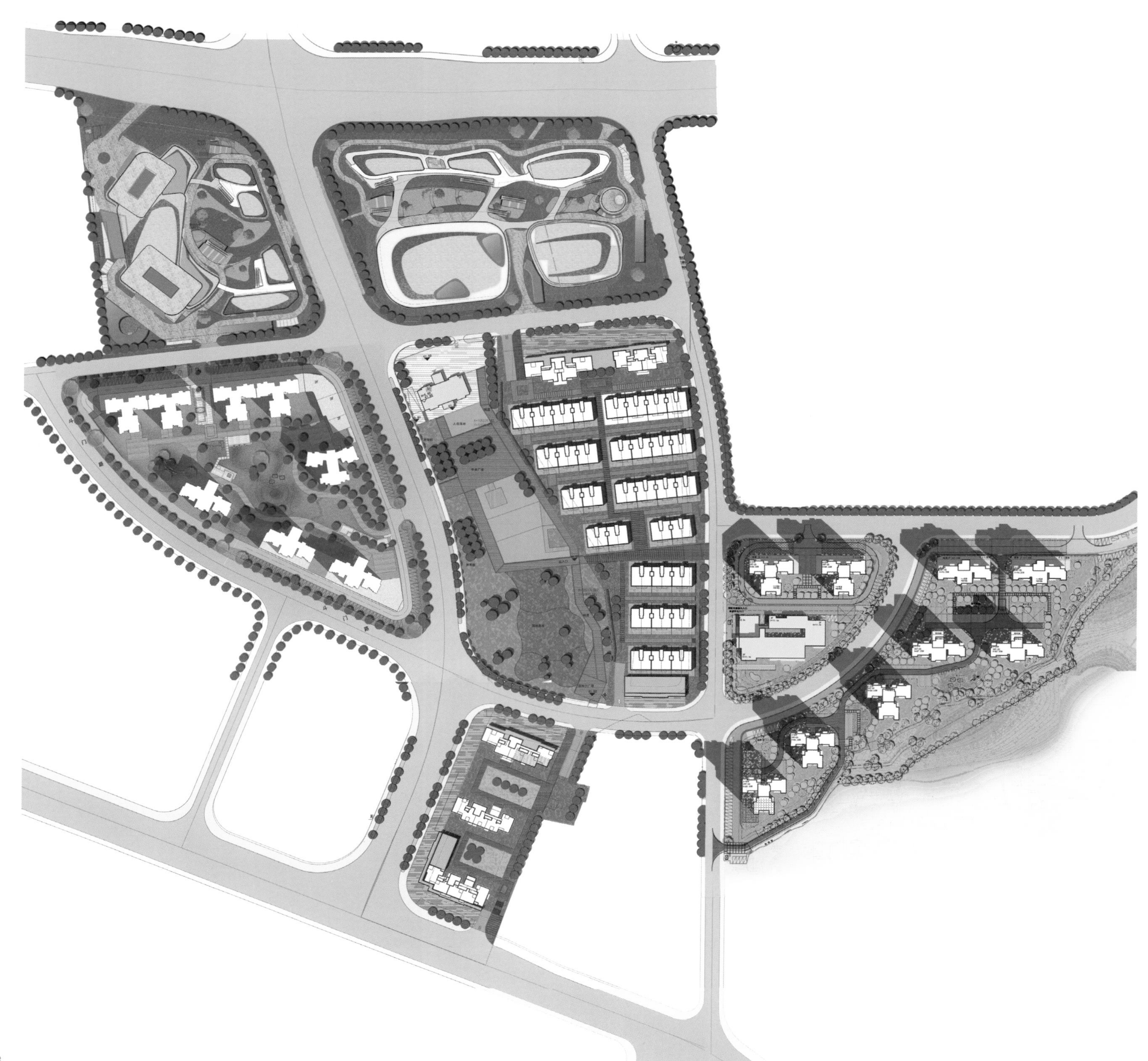

HUGO
Chanel
Chanel
Dior
AMONNI
Versace

CONIFER SOMO
SUBWAY
ForeBrand
ACURA
TEAMPORTAL

项目地理位置优越，集合资源优质，为长沙地区第一个流线型的街区商业综合体，也是五矿地产第一个商业综合体项目。项目定位为长沙下一代娱乐文化制作基地，提出“长沙 Live”的文娱主题，为城市提供创意性的娱乐空间。众多一线文化品牌进驻，赋予了该项目更多元的精神内涵。

项目整体形象干净时尚，以“菲林卷”为形态母体，演变出流动的建筑空间和立面形态。街区空间水平渗透，纵向互联。项目将三个独立的城市地块连接成形象统一、功能互补、流线互联的娱乐核心区。

With predominant location and collection of high-quality resources, the project is the first streamlined block-type commercial complex of Changsha and the first block-type commercial complex of Minmetals Real Estate. The positioning of the project is the next entertainment culture production base in Changsha, so the project puts forward the entertainment theme of “Changsha Live”, providing creative entertainment space for the city. Many of the first-tier cultural brands come here, giving the project more multiple spiritual contents.

The project, with clean and fashionable image, takes the “film roll” as the mother body, and evolves into a flowing architectural space and facade form. Block space permeates with each other horizontally and interconnects vertically.The project connects three independent urban blocks into an image-unified, function-complementary and streamline-connected entertainment core.

云南丽江雪山小镇艺术家工作室A

云南 丽江 | 2014

Snow Mountain Town Artist Studio A in Lijiang, Yunnan

Lijiang, Yunnan | 2014

性质：公共建筑
用地面积：362.50 m²
建筑面积：195.78 m²
开发商：丽江雪山投资有限责任公司
主题词：展览

Property: Public Architecture
Site Area: 362.50 m²
Floor Area: 195.78 m²
Client: Lijiang Xueshan Investment Co., Ltd.
Keywords: Exhibition

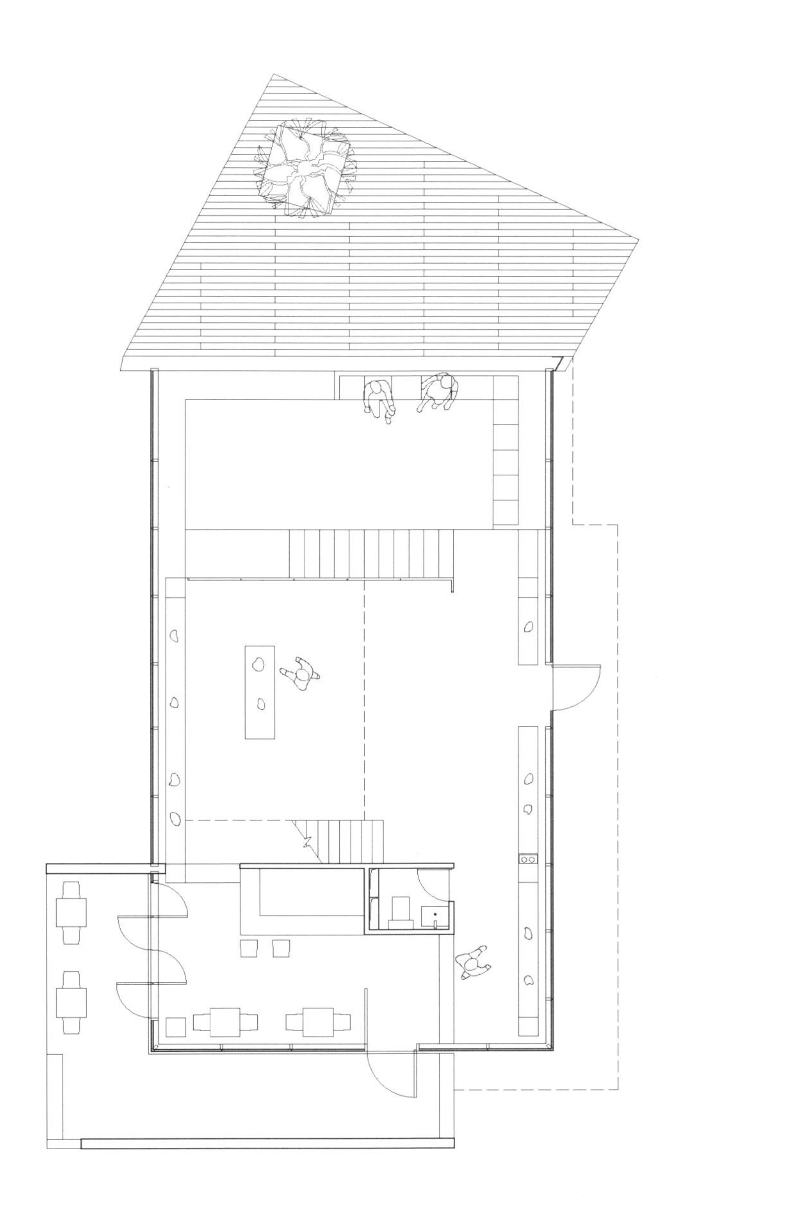

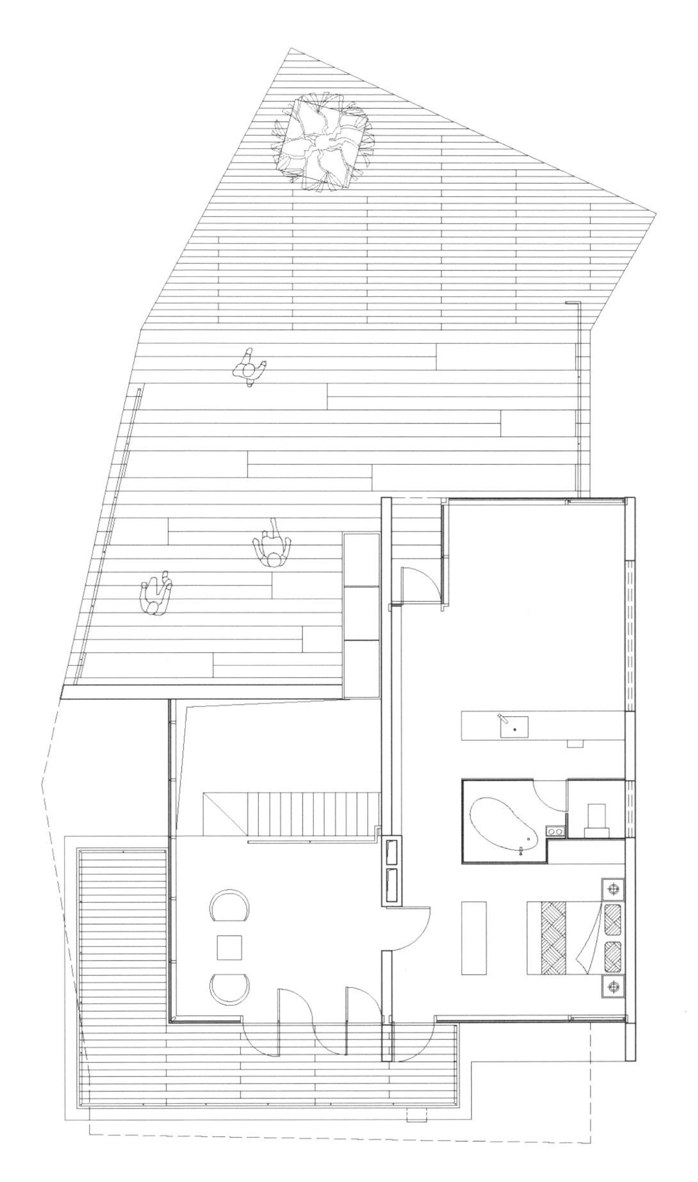

基地西、北临道路，北望玉龙雪山，视野开阔，位置优越。作为小型艺术论坛场所，建筑旨在积极融入场所，并进行友好干预，使之成为区域内的活跃因子。引用山形元素，由地面而上生成广场大台阶，削弱对街角的压迫感，同时将空间释放给公众，增加天然的交流场所。以似山的舒展形体面北，达到建筑与雪山、人与雪山积极对话的目的。方体的嵌入使形体更为丰富，玻璃的运用加强了室内外场所的互动。

The site has broad vision and privileged position with the roads on the west and north, and Yulong Snow Mountain to the north. As a small art forum, the building is designed to be actively integrated with the site and to engage in intervention friendly, making it an active factor in the region. The mountain elements are used to create square's large steps from the ground, giving more space to the public and increasing the natural communication places while reducing the pressure on the street corners. The mountain-like building faces the north, achieving the purpose of dialogue between the building and snow-mountain, and dialogue between people and the snow-mountain. The embedding of the square body makes the form richer, and the use of glass strengthens the interaction of indoor and outdoor places.

云南丽江雪山小镇艺术家工作室B

云南 丽江 | 2014

Snow Mountain Town Artist Studio B in Lijiang, Yunnan

Lijiang, Yunnan | 2014

性质：公共建筑
用地面积：200.00 m^2
建筑面积：202.54 m^2
开发商：丽江雪山投资有限责任公司
主题词：浮

Property: Public Architecture
Site Area: 200.00 m^2
Floor Area: 202.54 m^2
Client: Lijiang Xueshan Investment Co., Ltd.
Keywords: Floating

建筑功能以现代画创作为主。基地三面紧临周边建筑，仅北面开敞，用地较幽闭。建筑以基地无形的场所压力生成厚实形体，结合地域特色建造坡屋顶，引入柔和天光，使之成为光的容器。此外，以水注入灵气，将建筑托于水面之上。最后，结合露台空间打破部分坡屋顶，形成“取景框”，使之能北眺玉龙雪山。清水混凝土的基调令其更为内敛、质朴，静置其中或冥想或创作，自得其乐。

The main function of the building is for modern painting creation. The site is surrounded by buildings on three sides with only the northern side open. The overall body of the building is influenced by the invisible pressure of the surrounding buildings. The sloped roof which is constructed according to the local feature, allows ample light into interior space. The building on the water expresses an almost ethereal feeling. The patio space and partial roof form a viewing frame, where we can overlook Yulong Snow Mountain in the north. The concrete makes the building more reserved and simpler, and you will be happy when you stay here to meditate or create.

南宁华润置地广场

广西 南宁 | 2016

CR Land Plaza,Nanning

Nanning, Guangxi | 2016

性质：住宅
用地面积：72 319.00 m²
建筑面积：378 621.64 m²
开发商：南宁华润置地北湖房地产有限公司
主题词：超高层社区；还建；复合功能

Property: Residence
Site Area: 72,319.00 m²
Floor Area: 378,621.64 m²
Client: CR Land Co., Ltd.(Nanning)
Keywords: Super High-rise Community; Reconstruction; Compound Function

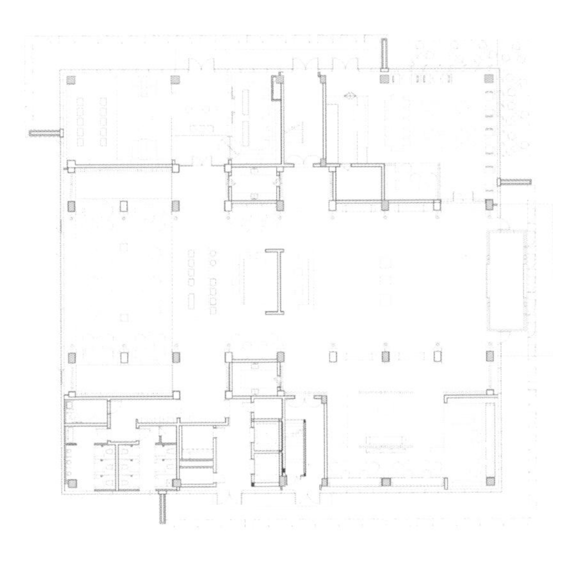

本项目是南宁市西乡塘区的重点旧改项目，依托华润集团强大的品牌优势和丰富的项目操作经验，打造集高端住宅小区、配套生活馆与公寓楼为一体的城市居住体，有力助推西乡塘区城市面貌的更新。

营销中心既是项目卖场，也是未来的社区邻里中心。设计改变了原来将泳池放在顶层的思路，经过多种方案比较，没有采取更具挑战的“桁构悬挑”方案。最终，泳池结合形体构成以“夹心饼干”样式呈现。

This project is the key reconstruction project of Xixiangtang District in Nanning City. Relying on strong brand advantages and rich experience in project operation of China Resources Group, the project creates an urban residence that integrates high-end residence area, supporting life hall and apartments , powerfully driving the renewal of Xixiangtang District.

The marketing center is both a project store and a future neighborhood center of community. The design changes the idea of placing the swimming pool on the top floor. There are many schemes in the process, but the more challenging scheme is not adopted. Finally, the pool is presented with the “sandwich biscuits”.

衡阳东路
北湖南路
明唐路
南棉路
明唐路

长沙华润桃源里

湖南 长沙 | 2016

CR Taoyuanli, Changsha

Changsha, Hunan | 2016

性质：住宅
用地面积：226 517.33 m^2
建筑面积：363 761.03 m^2
开发商：华润置地（湖南）发展有限公司
主题词：混合社区；创新别墅；新亚洲建筑风格；桃花源主题

Property:Residence
Site Area: 226,517.33 m^2
Floor Area: 363,761.03 m^2
Client: CR Land Co., Ltd.(Hunan)
Keywords: Mixed Community; Innovative Villa; New Asian Architectural Style; Theme of Peach Garden

TAO
YUAN
LI

项目依托周边得天独厚的自然景观环境，塑造多样化的小区空间和生活场所，以适应现代人多层次的需求。在前庭后院的基础上引入户内庭院的概念，将“一户三院”的创新别墅设计自然融入生活。典雅的新亚洲建筑风格，端庄大气。厚重的石材质感与通透的玻璃材质相对比，既显示了建筑的稳重，又不失现代感。深与浅的色彩对比，于古典中透露出现代的气息。

Relying on the unique natural landscape of the surrounding, the project creates a diversified residential space and living place to meet the multi-level needs of modern people. On the basis of the vestibular backyard, the concept of the indoor courtyard is introduced, and the innovative villa design of “one family and three courtyards” blends into life naturally. Elegant new Asian architectural style is dignified and decent. Compared with the transparent glass material, the thick stone texture shows the building’s steadiness and modern sense. The contrast of the deep and shallow color reveals the modern and classical style.

柳州华润凯旋门

广西 柳州 | 2014

CR Triumphal Arch, Liuzhou

Liuzhou, Guangxi | 2014

性质：公共建筑
用地面积：100 305.84 m²
建筑面积：504 484.78 m²
开发商：华润置地（柳州）有限公司
主题词：精细化；经济性；动感曲线

Property: Public Architecture
Site Area: 100,305.84 m²
Floor Area: 504,484.78 m²
Client: CR Land Co., Ltd. (Liuzhou)
Keywords: Elaboration; Economy; Dynamic Curve

销售中心兼顾前期项目展示及日后作为独立商业的功能需求，既是项目的公共休闲中心，也是城市道路三角洲地带的形象灯塔。

设计以经济性、可实施性与空间功能转换的适应性为基础，以流畅简洁的建筑语言打造具有动感的建筑形象。

建筑形体迎合人流线路，在南面主要出入口的位置适当退让，形成入口的形象广场和人流的集散地。项目规划布局高效合理，兼具了高品质和经济性的特质，最大化地利用了场地边界，布置活力商业环线，同时利用场地退线，创造局部街区商业。

超高层住宅形象统一，布局合理，每栋楼均有较好的观景角度，庭院通透度高。

The sales center takes into account the pre-project presentation and future functional requirements as an independent business, and is both a public leisurely center for the project and an image beacon for the urban road delta.

The design, based on the economy, implementation and adaptability of spatial function conversion, creates a dynamic architectural image with fluent and concise architectural language.

The building body caters to the human flow line, and provides a retreat at the south main entrance, which forms the image square of entrance and the distribution center of people. The project planning layout is efficient and reasonable, combines the characteristics of high quality and economy, maximizes the use of the site boundaries to arrange the dynamic business circle, and uses the site back line to create a part of block-style business.

 The image of super high-rise housing is unified; the layout is reasonable; each building has a good viewing angle; and the courtyard permeability is high.

南京金地自在城

江苏 南京 | 2009

Gemdale Megacity, Nanjing

Nanjing, Jiangsu | 2009

性质：住宅

用地面积：300 835 m²

建筑面积：590 191 m²

开发商：金地集团

主题词：早期现代主义风格；完成度的把握；混合社区

Property: Residence

Site Area: 300,835 m²

Floor Area: 590,191 m²

Client: Gemdale Corporation

Keywords: Early Modernism Style; Assurance of Completion; Mixed Community

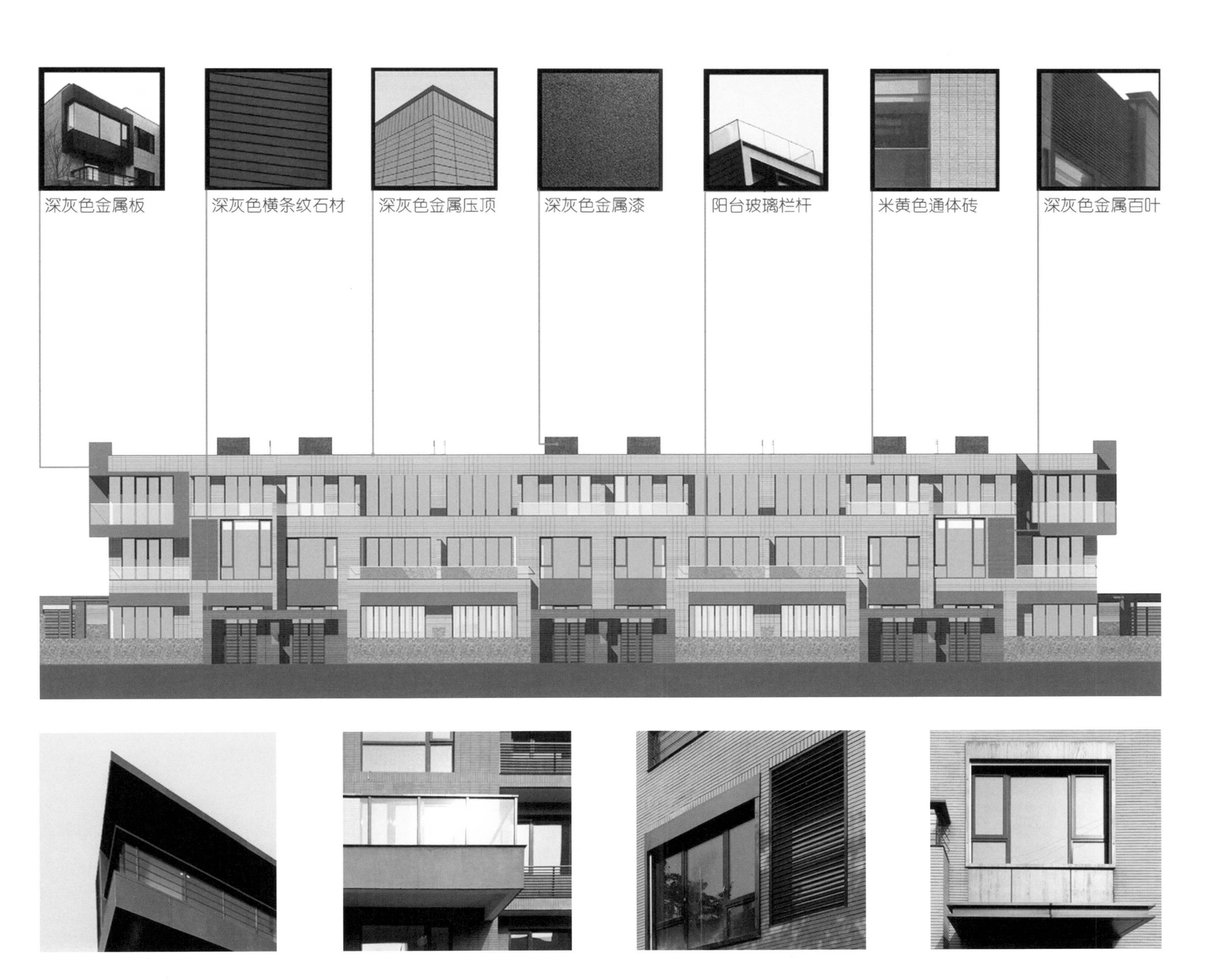

项目延续社区总体规划构思，打造东西向的社区生活主轴，并沿该主轴组织公共空间序列及设计流线，形成自身完整的社区系统。立面采用早期现代主义风格，选择符合建设成本要求的建筑工艺，并通过细节设计，有效实现了对完成度及设计品质的把握。

The project, involves the master planning of the Gemdale Megacity; creates east-west axis linking the communities together; and organizes the public space sequences and design streamline along the axis. The facade design adopts the early modernism style, the construction conforms to the requirements of construction cost, and the detail design effectively achieves the assurance of completion and design quality.

东莞东北师范大学附属益田小学

广东 东莞 | 2010

Yitian Primary School Attached to Northeast Normal University, Dongguan

Dongguan,Guangdong | 2010

性质：教育建筑

用地面积 : 31 810.49 m^2

建筑面积：30 683.49 m^2

开发商：益田集团

主题词：开放式教学；共享；自然；流动

Property: Educational Architecture

Site Area: 31,810.49 m^2

Floor Area: 30,683.49 m^2

Client: Yitian Group

Keywords: Open Teaching; Sharing; Nature; Flexibility

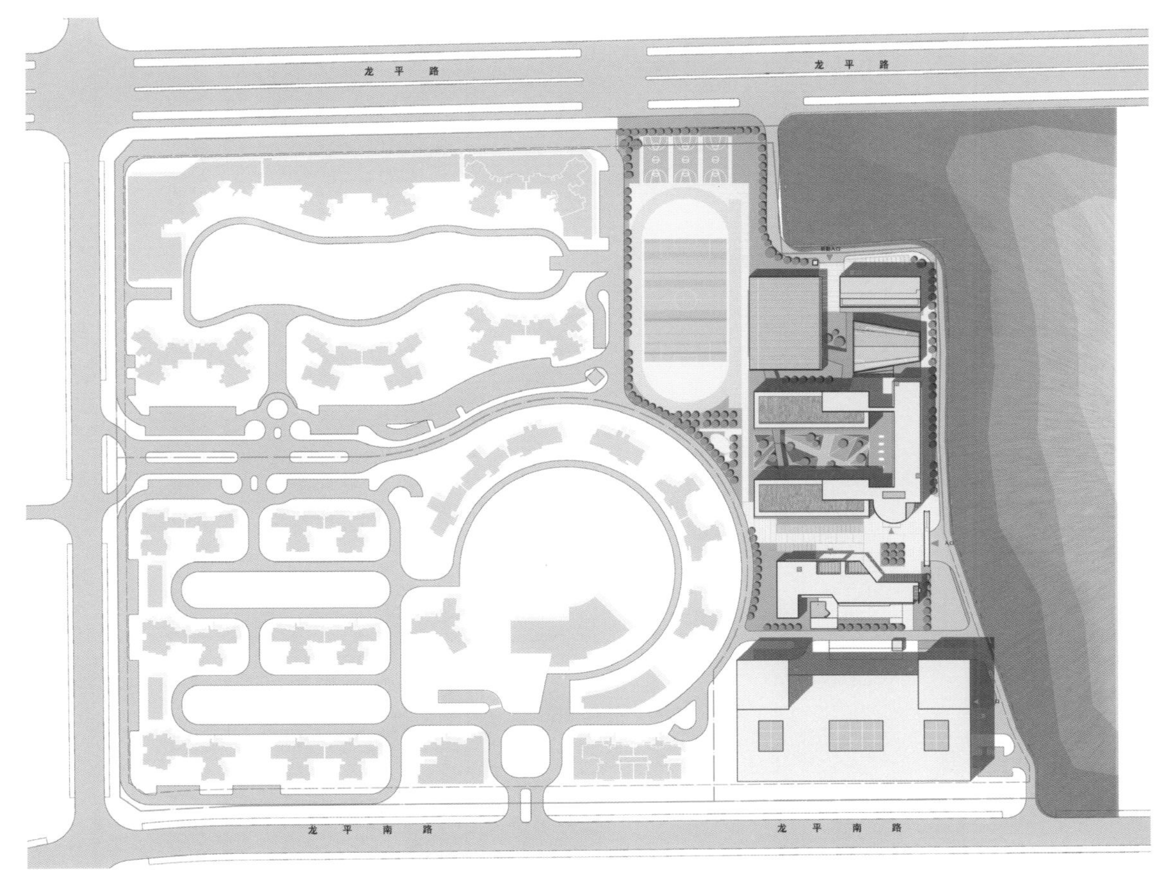

项目旨在创造自然温情且充满交流可能性的校园。以开放空间为核心，创造理想、适用的学习交流场所。空间丰富，强调室内外空间的开放、流动、过渡和共享。

立面设计继承简约的现代主义精神，兼顾自然质朴的形象。

The project aims to create a natural, warm and communicable campus. It emphasizes open spaces and creates an ideal and applicable learning space. The abundant space emphasizes openness, flexibility, transition and sharing from indoor to outdoor space.

The facade design adopts a concise modernist spirit, with an emphasis on natural and simple image of the architecture.

50×35×2厚
L形角钢连接件
木色弧形金属百叶
1.5厚

弧形百叶大样

白色涂料
屋顶
中灰色压型钢版（竖纹）
由专业厂家二次设计
耐候硅胶
室内
白色涂料
室内
灰色钢结构入口雨棚，尺寸详见平面
由专业公司负责设计、制作、安装

弧形玻璃幕墙

排水天沟≤300
中灰色压型钢版（竖纹）
由专业厂家二次设计
金属收边

压型钢板墙面剖面示意

白色涂料
屋顶
白色涂料
白色涂料
灰色穿孔铝板
白色涂料
室内

教室外墙剖面示意

50×35×2厚
L形角钢连接件
木色弧形金属百叶
1.5厚

弧形百叶背立面

0.77mm 镀锌压型钢板 中灰色（亚光）

压型钢板水平断面示意

金属百叶
螺栓固定
L形角钢连接件
80mm场∠50×35×5
预埋件

深圳坪山锦绣学校

广东 深圳 | 2018

Pingshan Jinxiu Primary and Secondary School, Shenzhen

Shenzhen, Guangdong | 2018

性质：教育建筑
用地面积：24 400.00 m²
建筑面积：85 531.73 m²
开发商：华润置地（深圳）有限公司
主题词：环

Property: Educational Architecture
Site Area: 24,400.00 m²
Floor Area: 85,531.73 m²
Client: CR Land Co., Ltd. (Shenzhen)
Keywords: Ring

沿退红线设计的环状路径建筑，最大化利用场地周长，从而实现了“平均最小进深”的空间效果，以及双面采光和通风的最大化。

建筑体量采用环绕形式设计，围合出中央花园广场，无论将来周边开发强度如何，都保证了建筑景观视野的最大化。

The circular path building along the red line is designed to maximize the use of the perimeter of the site, thus achieving the space effect of “average minimum depth of entry” and maximizing the double-sided lighting and ventilation.

The building is designed in the form of ring and encloses a central garden square. Regardless of the intensity of the future development, it ensures the maximization of the landscape view of the building.

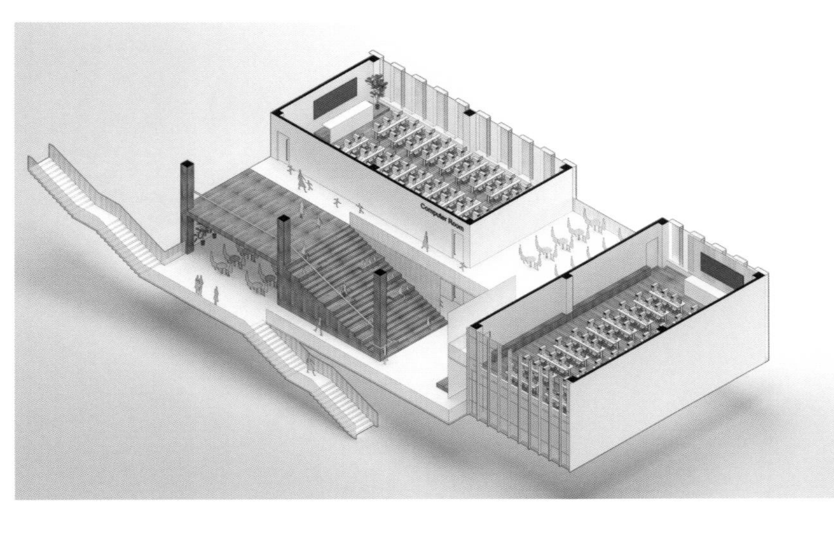

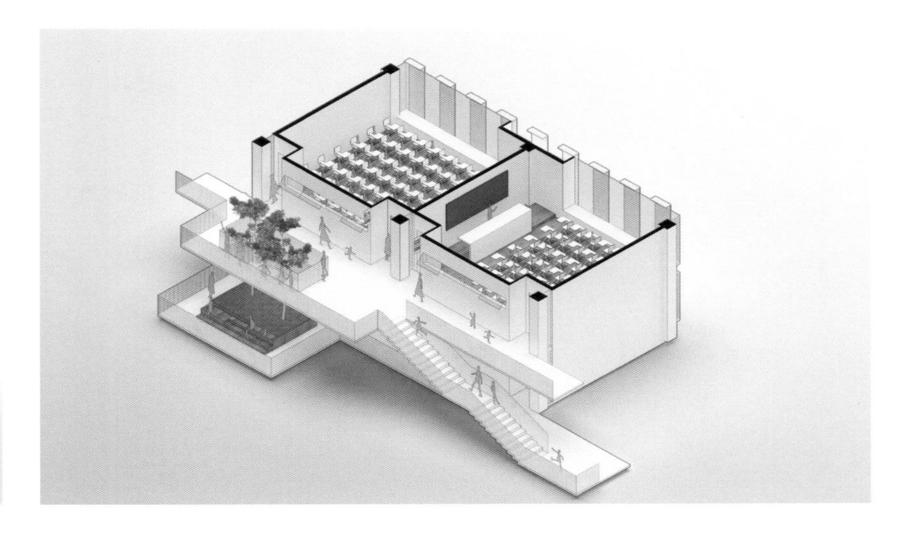

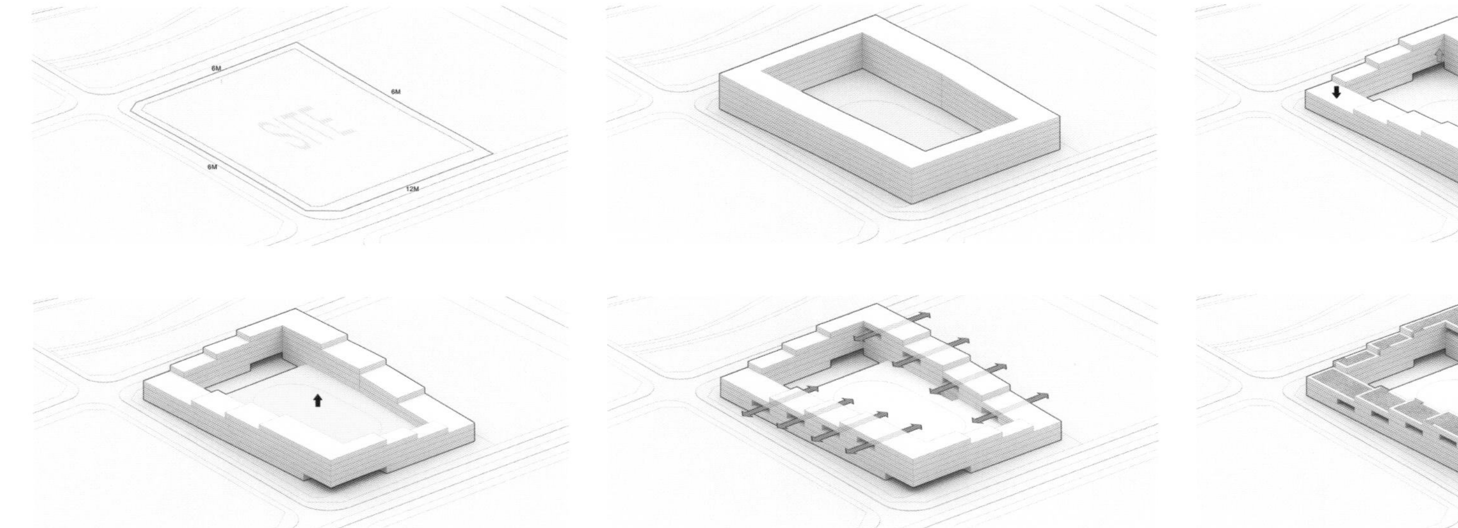

退台式的建筑形态，使得从东、南、西、北四个方向观看，都可以获得有空间进深感的景观视角，实现了“远观”的观景体验。同时，退台的方向考虑了东南主导风向，帮助场地内更好地通风和采光。

项目自下而上功能优越，自上而下形象统一。建筑形态舒缓低矮，兼顾整体和独特的视觉效果，可辨识性高。精心设计的穿孔铝板建筑表皮，在模数的控制下，可实现三种立面效果，既统一又丰富。

The architectural form of the terrace makes it possible to provide a view of the landscape with spatial depth from the four directions of east, south, west and north, and to have the viewing experience of “distant view” . At the same time, the direction of the terrace takes into account the prevailing wind direction from the southeast to have better ventilation and lighting in the area.

The project has superior bottom-up functions and a unified image from top to bottom. The architectural form is low and smoothing, and it has an overall and unique visual effect and is highly recognizable. Under the control of modulus, carefully designed building skin of perforated aluminum sheet can have three kinds of facade effects, both unified and rich.

深圳华润九年制义务学校

广东 深圳 | 2014

CR Nine - year Compulsory School, Shenzhen

Shenzhen, Guangdong | 2014

性质：教育建筑
用地面积：35 655 m²
建筑面积：41 896 m²
开发商：华润置地（深圳）有限公司
主题词：契合；围合；整合

Property: Educational Architecture
Site Area: 35,655 m²
Floor Area: 41,896 m²
Client: CR Land Co., Ltd. (Shenzhen)
Keywords: Fitting; Enclosure; Integration

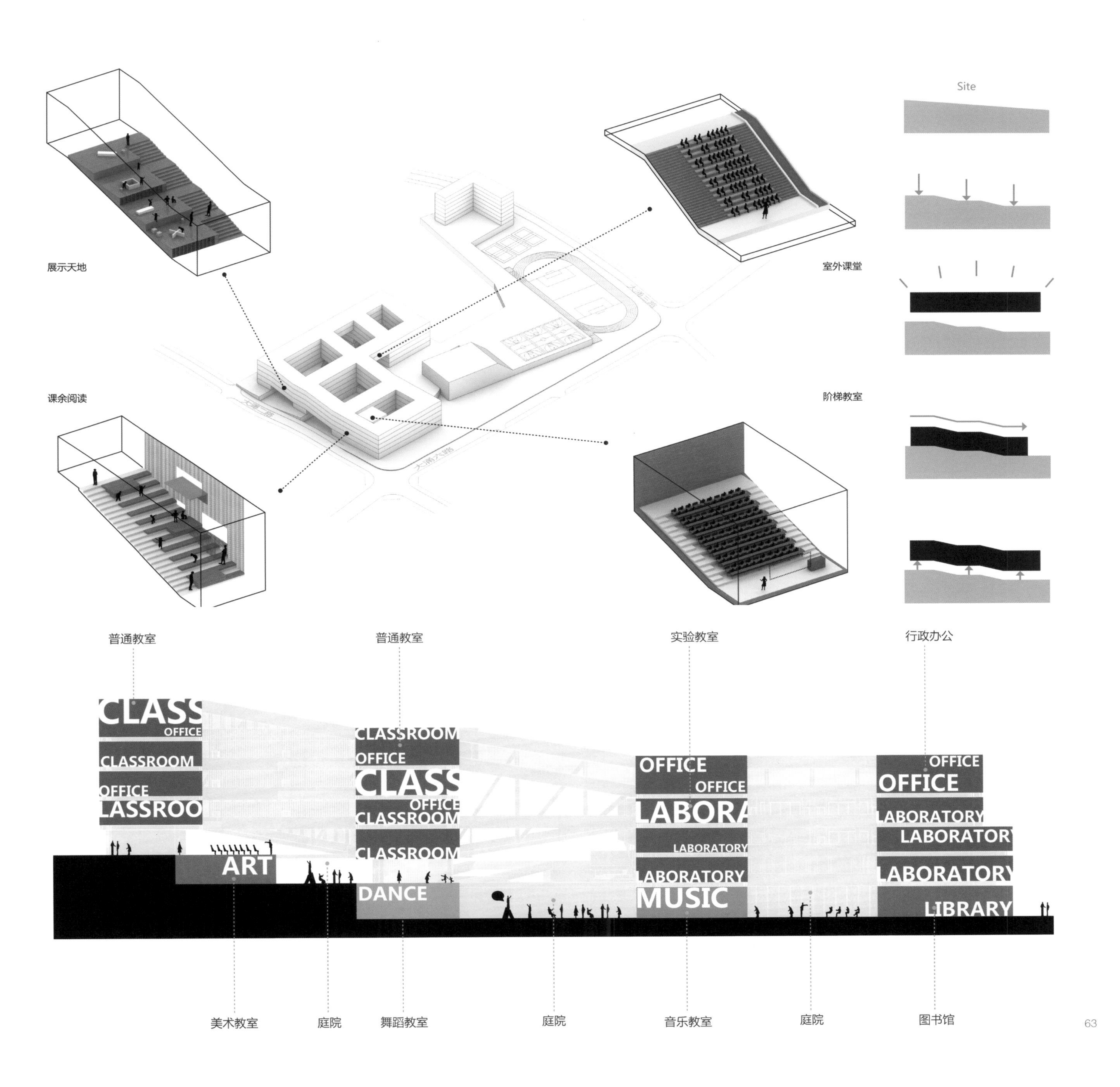

Site
展示天地
室外课堂
课余阅读
阶梯教室
普通教室
普通教室
实验教室
行政办公
CLASS
OFFICE
CLASSROOM
OFFICE
LASSROO
CLASSROOM
OFFICE
CLASS
OFFICE
CLASSROOM
CLASSROOM
OFFICE
OFFICE
LABORA
LABORATORY
LABORATORY
MUSIC
OFFICE
OFFICE
LABORATORY
LABORATORY
LABORATORY
LIBRARY
ART
DANCE
美术教室
庭院
舞蹈教室
庭院
音乐教室
庭院
图书馆

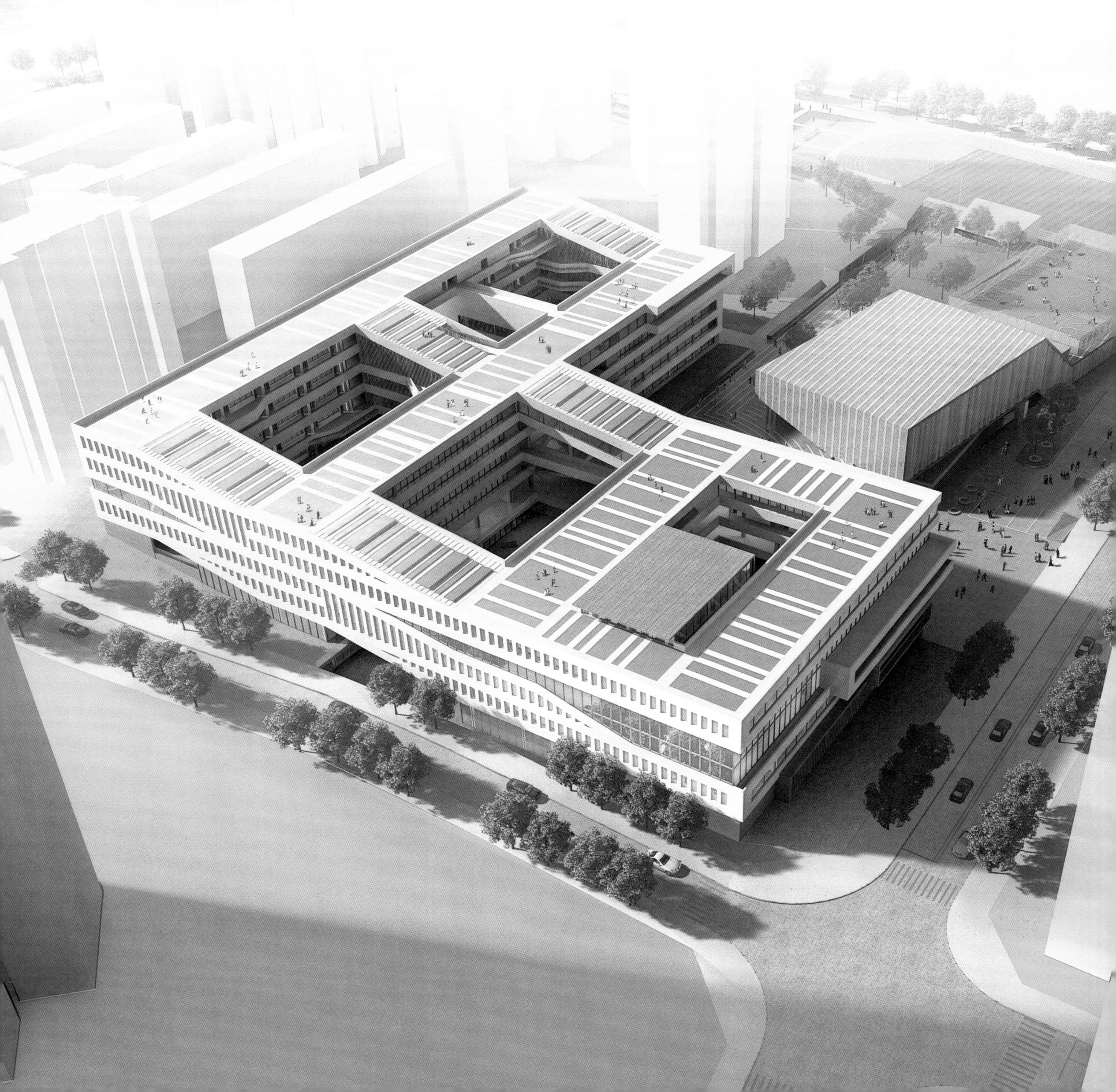

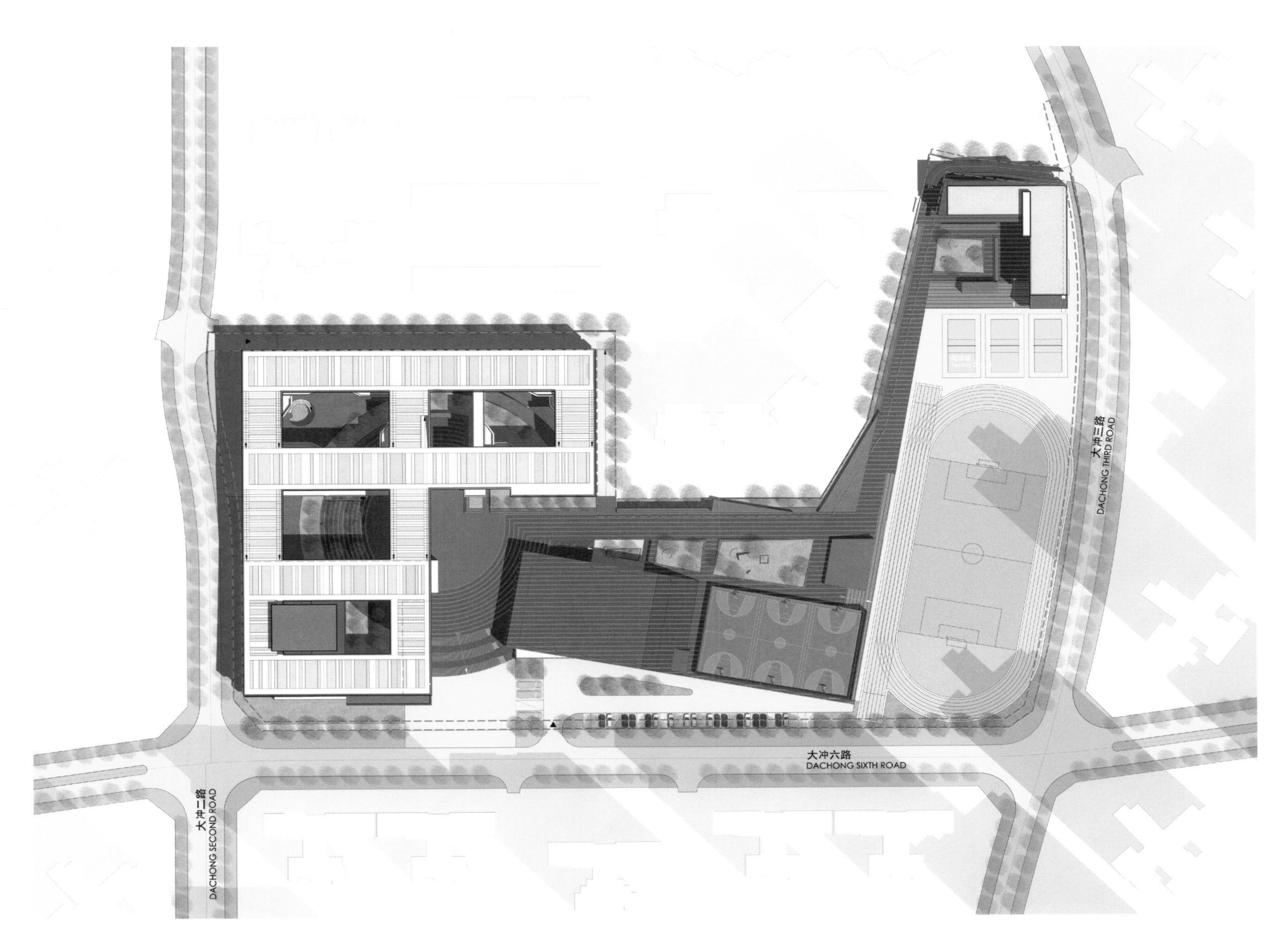

项目主体建筑集中设置在场地的东西两侧，两者围合出中部的开阔绿地，其间可布置风雨操场、运动场地和公园。

教育轴和休闲轴交织的张力及其各自包含的丰富功能，造就了令人惊叹的空间复杂性，使得这个功能繁复的校园建筑具备了城市性的体验。

The main buildings of the project are concentrated to locate on both sides of the site. The two sides enclose an open green space in the middle, where the sports ground with roof, the playground and the park can be arranged.

The tension between the educational axis and leisurely axis, and the rich functions create surprising spatial complexity, which makes this complex campus building possess a kind of urban experience.

深圳白石洲村综合整治

广东 深圳 | 2018

Comprehensive Renovation of Baishizhou Village, Shenzhen

Shenzhen, Guangdong | 2018

性质：旧村综合整治
用地面积 : 160 000 m^2
建筑面积：480 000 m^2
开发商：华润置地（深圳）有限公司
主题词：城市更新；街区活化；旧村整治；人才小镇

Property: Comprehensive Old Village Renovation
Site Area: 160,000 m^2
Floor Area: 480,000 m^2
Client: CR Land Co., Ltd. (Shenzhen)
Keywords: Urban Renewal; Neighborhood Activation; Old Village Renovation; Talent Town

中原地产
CENTALINE PROPERTY
7天连锁酒店

白石洲

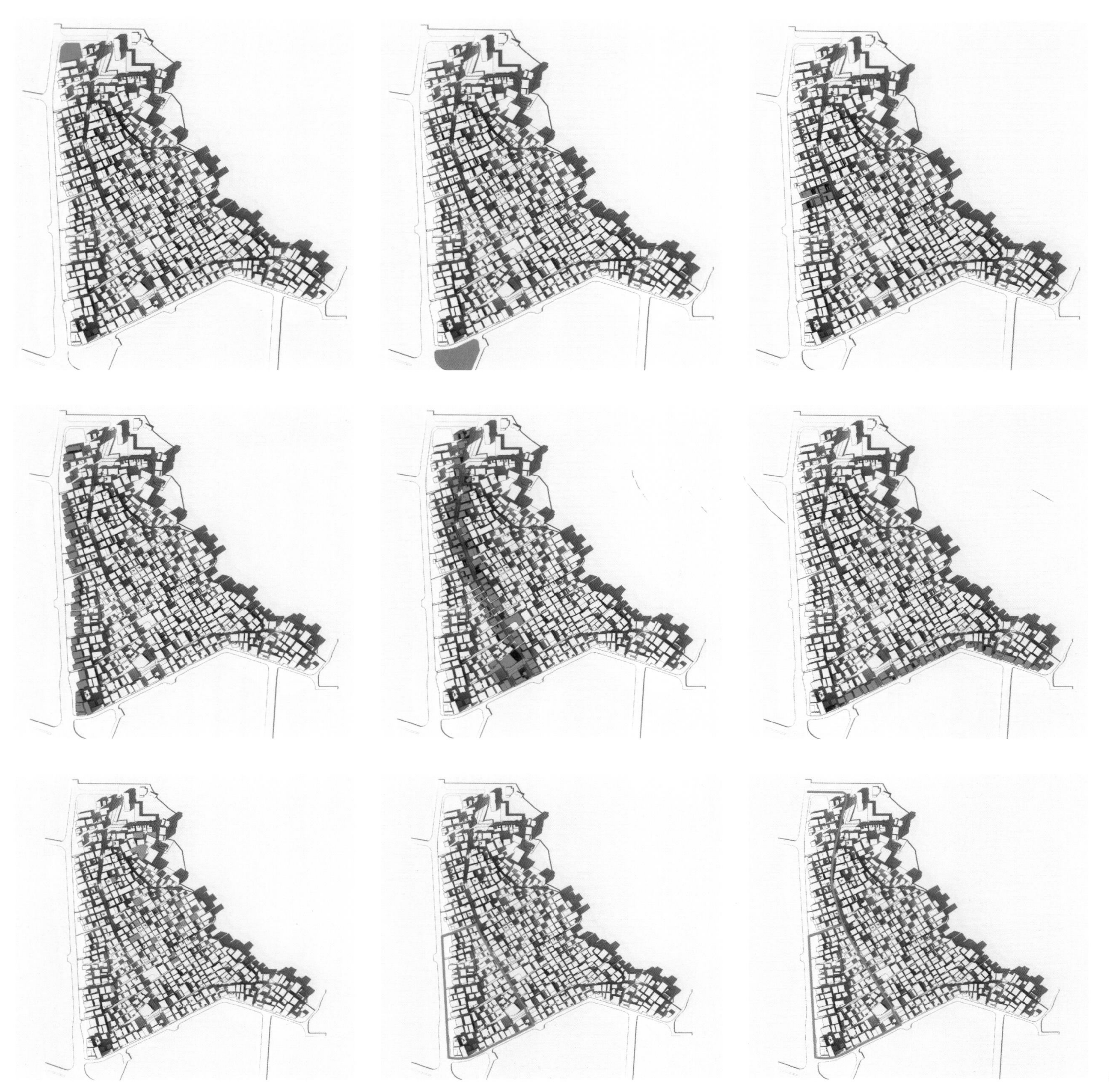

白石洲村位于深南大道边，是深圳最大的几个“城中村”之一。它拥有深圳市区几乎最集中、最大规模的农民房，绝大部分从外地来深圳的人都有过一段或长或短的在白石洲生活的记忆。

在改善居民基本生活工作条件的基础上，白石洲村的综合整治以《深圳市“城中村”综合治理行动计划（2018—2020年）》为指导，通过公共空间设计、立面设计、道路景观设计、VI系统设计等方法，为城市带来更有文化内涵和建筑品味的街道空间，从而实现提高生活质量、提升城市形象、吸引人才、留住人才、激发城市活力的目标。

Located on Shennan Avenue, Baishizhou Village is one of the largest “villages in the city” of Shenzhen. It has almost the most concentrated largest peasant houses in Shenzhen. Most of the people who come to Shenzhen from other places have a long or short life in Baishizhou.

Based on the basic conditions for improving the living and working conditions of residents, guided by the “The Comprehensive Governance Action Plan for ‘Village in the City’ (2018–2020)”, and through public space design, facade design, road landscape design, VI system design and other methods, comprehensive renovation of Baishizhou Village brings street space of more cultural connotation and architectural taste to the city, thus achieving the purpose of improving the quality of life, enhancing the image of city, introducing talents, retaining talents and stimulating the vitality of the city.

东莞益田假日天地

广东 东莞 | 2010

Yitian Holiday Palace, Dongguan

Dongguan,Guangdong | 2010

性质：商业综合体
用地面积：61 486 m²
建筑面积：153 385 m²
开发商：益田集团
概念设计合作方：加拿大Forrec
主题词：娱乐主题商场；街区商业；多元动感

Property: Commercial Complex
Site Area: 61,486 m²
Floor Area: 153,385 m²
Client: Yitian Group
Cooperating Unit: Canada Forrec
Keywords: Entertainment-theme Mall; Block Business; Multiple and Dynamic

项目旨在打造邻里尺度的商业综合体以及拥有商业氛围的城市游乐地，在多元化业态经营的同时，着力增加娱乐主题不同的室内游乐项目，使这里成为人们购物、休闲、娱乐的最佳去处。东西向的弧线内街将综合体分为两部分，在化解庞大体量的同时，营造了动感的外部空间。中庭各节点空间充满变化，消费者徜徉其中，一步一景，妙趣横生。

The project aims to build a neighborhood-scale commercial complex as well as a civic recreational area with the commercial atmosphere. Multi-scheme commercial buildings provide a wide range of indoor recreational activities, making it the best place to shop, relax and entertain. The curved inner street from the east to west divides the commercial complex into two parts. Therefore,while resolving the large buildings volume, it creates a dynamic exterior space. Each atrium space is full of various changes, which creates a more enjoyable experience for the public.

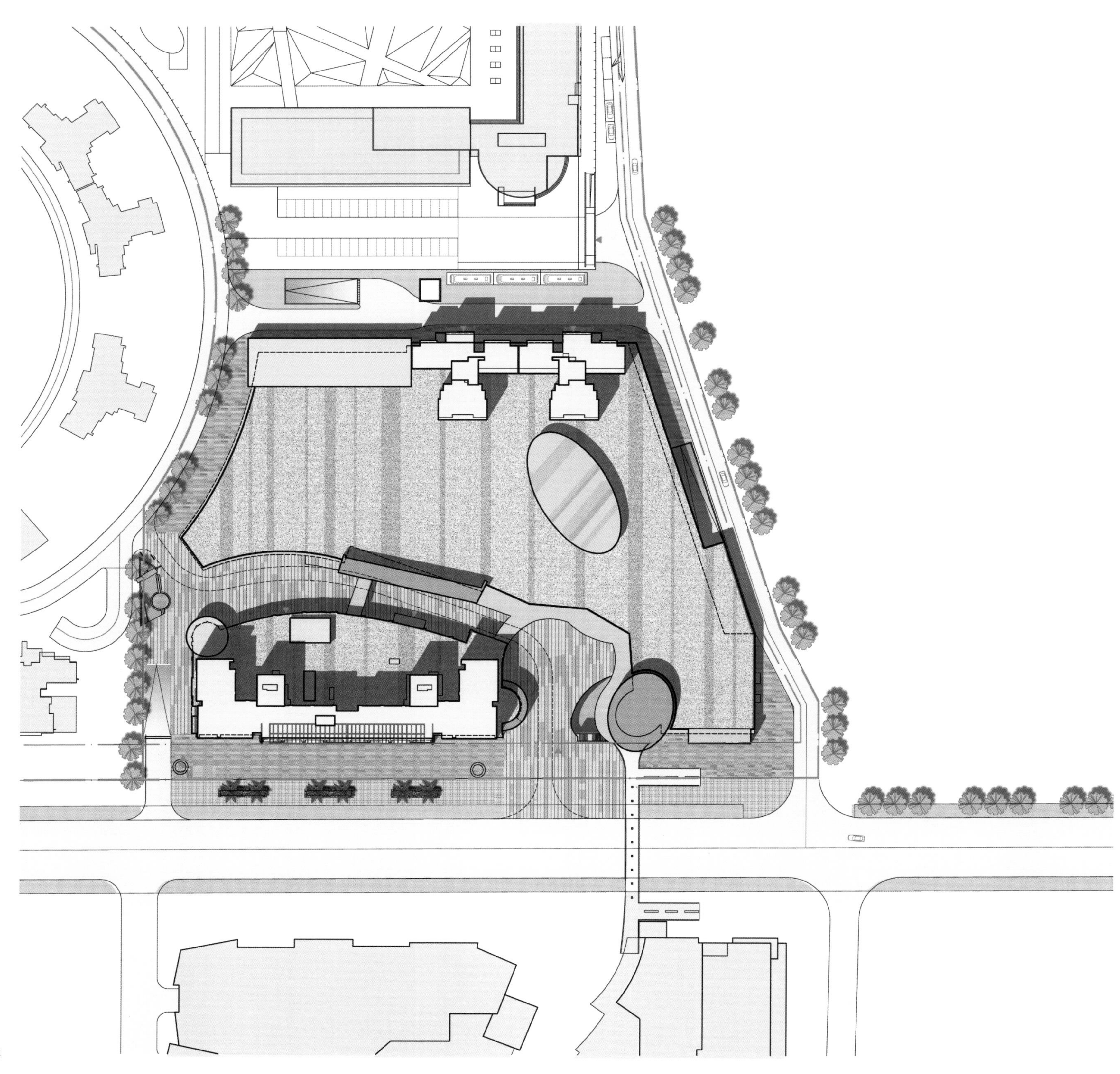

aloft
HOTELS
W xyz lounge

重庆华润二十四城万象里

重庆 | 2012

CR MIXC of Twenty-four City, Chongqing

Chongqing | 2012

性质：商业综合体
用地面积：27 765 m²
建筑面积：192 588 m²
开发商：华润置地（重庆）有限公司
主题词：销售型街区商业；混合业态；核心商圈

Property: Commercial Complex
Site Area: 27,765 m²
Floor Area: 192,588 m²
Client: CR Land Co., Ltd. (Chongqing)
Keywords: Sales-oriented Block Business; Mixed-use Business; Commercial Center

万象繁华 火热招商
自助扫码购
任性不排队

项目毗邻重庆华润万象城，作为城市核心商圈的延续，利用Lifestyle商业模式为城市提供开放舒适的商业及休闲空间。项目引入了混杂的商业业态，方案设计融合了购物中心、街区商业、地下超市及商业内街等多种商业空间，并运用下沉广场、退台、连廊等建筑手法打造丰富多元的街区商业综合体。

Close to the Chongqing MIXC, the project as a continuation of the city commercial center, introduces a Lifestyle business model to create an open and comfortable commercial and leisurely space. The project includes the mixed-use business of shopping mall, commercial street, supermarket, commercial inner street and other commercial types. The design creates a diversified block business complex by architectural techniques such as sunk plaza, terrace and corridor.

浏阳中能建中央国际广场

湖南 浏阳 | 2015

CECH Central International Square, Liuyang

Liuyang, Hunan | 2015

性质：商业综合体
用地面积：69 060 m²
建筑面积：290 466 m²
开发商：湖南鼎鹰房地产开发有限公司
主题词：城市中心商业；街区商业；缝合与织补

Property: Commercial Complex
Site Area: 69,060 m²
Floor Area: 290,466 m²
Client: Hunan Dingying Real Estate Development Co., Ltd.
Keywords: City Center Business; Block Business; Stitching and Weaving

销售中心
Liuyang Sales Center

S I S L E Y S

GAP

110KV高压走廊
劳
动
路
北
正
北
路

项目基地地处浏阳市老城区，文化积淀深厚，优势明显，但周边环境比较复杂，公共配套设施不够完备。项目依托基地周边优势环境，以街区商业模式连接原先割裂的城市地块，重新激活本应存在的城市核心，为该地区带来时尚、科技、丰富的商业空间。在空间组织上，合理利用场地高差，以公共街区为线索，串联各个标高的建筑空间。通过交通设计，有效管理住宅生活区和商业活力区的边界，实现高效而安全的城市生活。

The base of the project,located in the old urban area of Liuyang City, has deep cultural accumulation and obvious advantages. However, the surrounding environment is relatively complex and the public facilities are not complete. Relying on the advantageous environment around the base,the project adopts the block business model to connect the original fragmented urban plots, thus reactivating the city center which should have existed and bringing fashion, technology and rich commercial space to the region. In regard to the space organization, it utilizes the difference of site height reasonably and connects the building space by taking the public block as the clue. Through traffic design, it effectively manages the boundaries of residential living areas and business vitality zones to realize efficient and safe urban life.

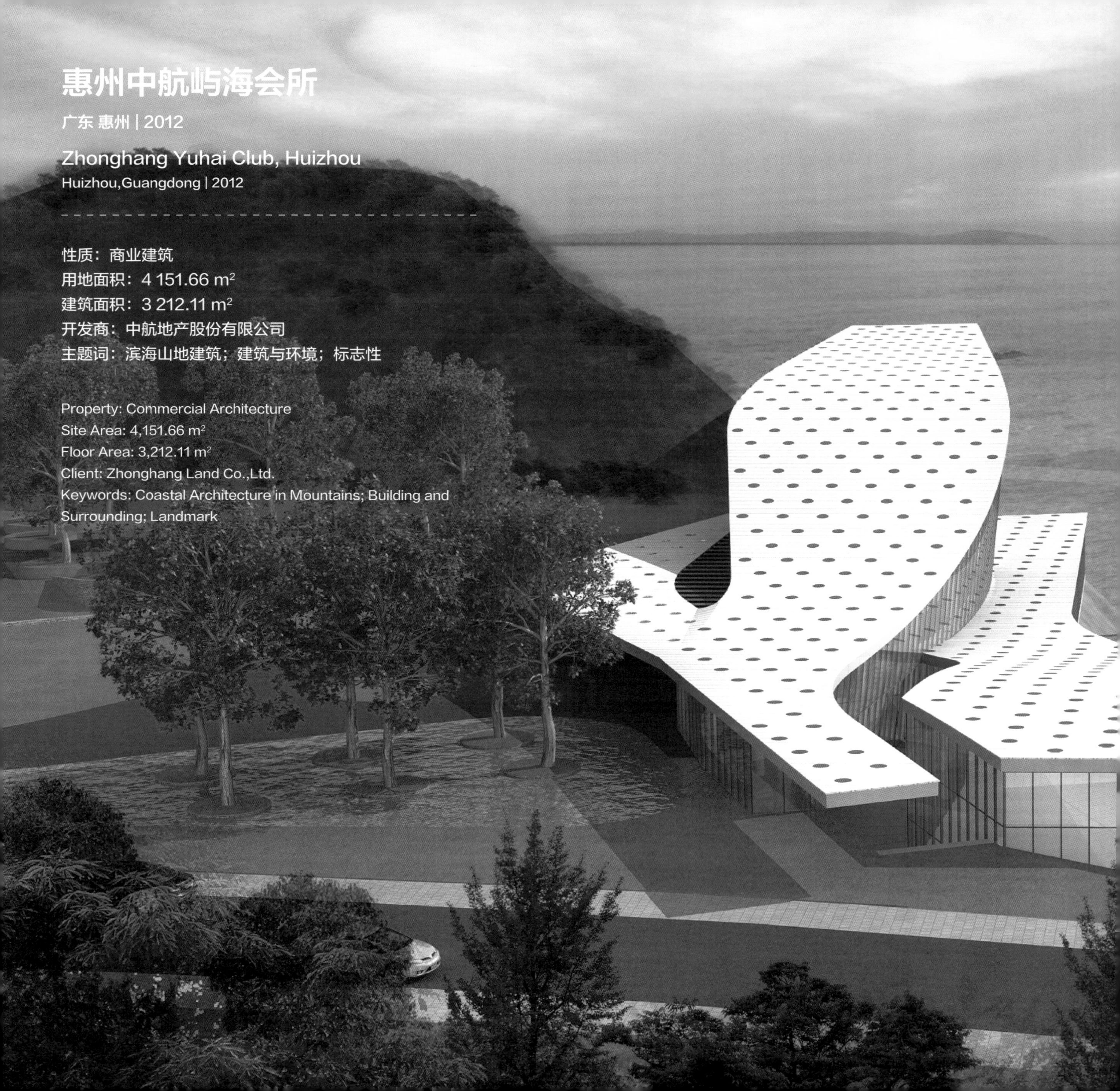

惠州中航屿海会所

广东 惠州 | 2012

Zhonghang Yuhai Club, Huizhou

Huizhou,Guangdong | 2012

性质：商业建筑
用地面积：4 151.66 m²
建筑面积：3 212.11 m²
开发商：中航地产股份有限公司
主题词：滨海山地建筑；建筑与环境；标志性

Property: Commercial Architecture
Site Area: 4,151.66 m²
Floor Area: 3,212.11 m²
Client: Zhonghang Land Co.,Ltd.
Keywords: Coastal Architecture in Mountains; Building and Surrounding; Landmark

会所设计方案注重建筑自身与山、海三者之间的总体关系，取意于富有海滨特色的舰船、海岛形态，以极具雕塑感的形态打造标志性的建筑，营造奢华的建筑体验。

The club design considers the relationship between the building, mountain and sea as one of the most important guidelines of this project. It uses the metaphor of ship and island to create a sculptural landmark architecture which creates a distinguished luxury experience.

惠州中航屿海观景平台

广东 惠州 | 2012

Zhonghang Yuhai Viewing Platform, Huizhou

Huizhou, Guangdong | 2012

性质：景观
用地面积：3 700.00 m²
建筑面积：2 955.16 m²
开发商：中航地产股份有限公司
主题词：建筑与景观；竖向设计；场所营造

Property: Landscape
Site Area: 3,700.00 m²
Floor Area: 2,955.16 m²
Client: Zhonghang Land Co.,Ltd.
Keywords: Architecture and Landscape; Vertical Design; Site Construction

景观栈道设有观景平台、绿植、水景、坡地，花园的屋顶与过街涵洞出口平台构成一个循环结构的连续景观，共同形成一片建筑化景观。

The landscape trestle road has a viewing platform, green planting, water view and slope, and the roof of the garden and the exit platform of the street culvert form a continuous landscape with a circular structure, forming an architectural landscape.

吉安庐陵文化城

江西 吉安 | 2018

Luling Cultural City, Ji’an

Ji’an, Jiangxi | 2018

性质：公共建筑
用地面积：32 900 m²
建筑面积：20 960 m²
开发商：吉安城市建设投资开发公司
主题词：地景；轴线；对称

Property: Public Architecture
Site Area: 32,900 m²
Floor Area: 20,960 m²
Client: Ji’an City Investment Company
Keywords: Landscape; Axis; Symmetry

吉安庐陵文化城项目基于已有的圆形阵列桩基，重新对整体建筑形态及空间进行优化设计。在尽量沿用已有桩基的基础上，本设计注重场地轴线的连贯性和流线的导向性，以双鱼玉佩为原型，创造出对称且不呆板的地景式建筑。建筑形态舒缓自然，屋顶及室内皆可自由到达。场地轴线从两座半月形建筑物中间穿过，与建筑融为一体。

Based on the existing circular array pile foundation, the project of Luling Cultural City in Ji'an redesigned the overall architectural form and space. Ensuring that the existing pile foundation is used as much as possible, the design focuses on the continuity and streamline orientation of the site axes, and uses the double-fish jade pendant as the prototype to create the symmetrical and non-stubborn landscape architecture. The form of the building is smooth and natural, and the roof and the interior are freely accessible. The axis of the site passes through the middle of the two half-moon buildings, combing with the building.

北京富根大厦

北京 | 2012

Fugen Tower, Beijing

Beijing | 2012

性质：办公楼
用地面积：9 213 m²
建筑面积：64 131 m²
开发商：北京福星晓程电子科技股份有限公司
主题词：功能混搭；现代简约

Property: Office Building
Site Area: 9,213 m²
Floor Area: 64,131 m²
Client: Beijing Fuxing Xiaocheng Electronics Co., Ltd.
Keywords: Mixed Functions; Modern and Concise

项目坐落在西长安街，并应客户要求建造融合商业、办公、公寓等多种业态的综合性办公楼。建筑布局因地制宜，在规划要求及用地条件的各种限制中开展设计，以高效舒适、节能环保的空间体验，简约大气、庄重典雅的现代气质，打造人性化的办公空间。

The project is a comprehensive office building located at west Chang'an Street, and contains commercial office and apartment sections upon the client's request. The general layout is adjusted according to local conditions, policies and restrictions. The design aims to create a humanized working space to get the efficient, comfortable and energy-saving space experience, which is full of simple and elegant temperament.

深圳迪富宾馆片区城市设计

广东 深圳 | 2012

Difu Hotel Area Urban Design, Shenzhen

Shenzhen, Guangdong | 2012

性质：综合性建筑
用地面积：16 655 m^2
建筑面积：177 944.4 m^2
开发商：深圳桑达电子集团有限公司
主题词：城市更新；商住综合体

Property: Complex Architecture
Site Area: 16,655 m^2
Floor Area: 177,944.4 m^2
Client: Shenzhen SED Electronics Group Co.,Ltd.
Keywords: Urban Renewal; Commercial and Residential Complex

基于企业性质和华强北电子商业区特点，建筑立面表皮采用一种数码化的肌理，从而有效地对基地文化做出回应。这层“数码表皮”有效地将四栋建筑及裙房外围联结成一个具有标志性和地域特色的整体。四栋塔楼内侧表面则采用通透纯粹的玻璃幕墙，这样可以使内部空间更加开阔通透。“数码表皮”的设计考虑了自上而下的呈现由密到疏的肌理关系，巧妙地与建筑内不同功能空间的采光需求相呼应。而实体墙加落地窗的方式既节能又有效地控制了造价。

Based on the property of enterprises and the characteristics of North Huaqiang Electronic Business District, the skin of building facade adopts a kind of digital texture, thus effectively responding to the culture of the base. The “digital skin” effectively connects the four buildings and the periphery of the podium into a whole with symbolic and regional feature. The inside surface of the four buildings is made of transparent glass curtain walls, which can make the inner space more open and transparent. The design of “digital skin” takes into account the texture of the top-down representation from dense to sparse, which is ingenious in response to the lighting needs of different functional spaces in the building. The use of solid wall and French window can save energy and control cost effectively.

成都华润凤凰城营销中心

四川 成都 | 2007

Sales Center of CR Phoenix City, Chengdu

Chengdu, Sichuan | 2007

性质：公共建筑
用地面积：530 m^2
建筑面积：215 m^2
开发商：华润置地（成都）有限公司
主题词：旧建筑改造；表皮；现代；时尚

Property: Public Architecture
Site Area: 530 m^2
Floor Area: 215 m^2
Client: CR Land Co., Ltd. (Chengdu)
Keywords: Renovation of Old Buildings; Skin; Modern; Fashion

本项目将原有小书店进行改造，设计上保留原来的墙体及结构，赋予其极具现代感的建筑表皮。运用横向及竖向的钢构架，构成了富于虚实变化的立面形式。镜面反射的不锈钢构件简洁时尚，也使得建筑和周围景观相映成趣，引人驻足。

The project aims to renovate the existing small bookshop. The design retains the structural elements and adds a modern skin to the structure. The design uses horizontal steel frames that wrap around vertical steel frames to compose unique facades. The glossy reflective stainless steel components look concise and fashionable, making the building be in harmony with the surrounding landscape, which is so charming to pedestrian.

华润置地
CR Land
与城市的未来一起飞翔
新会展 · 纯居住 · 艺术社区
new exhibition · pure living · artistic community
项目地址：成都市新会展片区站华路
028-8125 9999
开发商：华润置地（成都）有限公司
China Resources Land Co.,Ltd. (Chengdu)
• 公司网址Http: //cd.crland.com.hk
凤凰

凤凰城销售中心
PHOENIX CITY
凤凰城

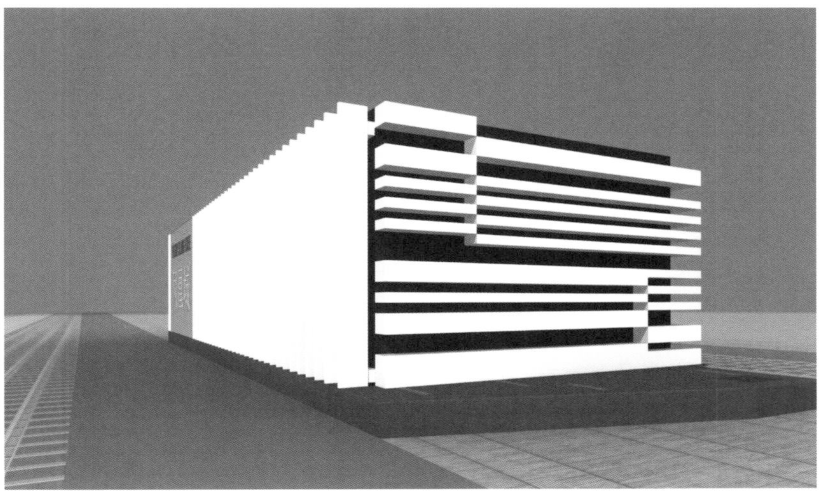

凤凰城销售中心
PHOENIX CITY
凤凰城

凤凰城

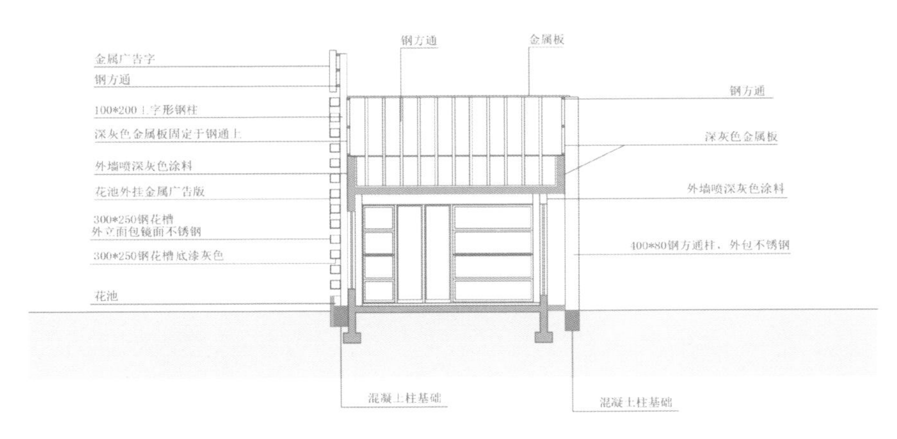
金属广告字
钢方通
100*200工字形钢柱
深灰色金属板固定于钢通上
外墙喷深灰色涂料
花池外挂金属广告版
300*250钢花槽
外立面包镜面不锈钢
300*250钢花槽底漆灰色
花池
钢方通
金属板
钢方通
深灰色金属板
外墙喷深灰色涂料
400*80钢方通杆，外包不锈钢
混凝上柱基础
混凝土柱基础

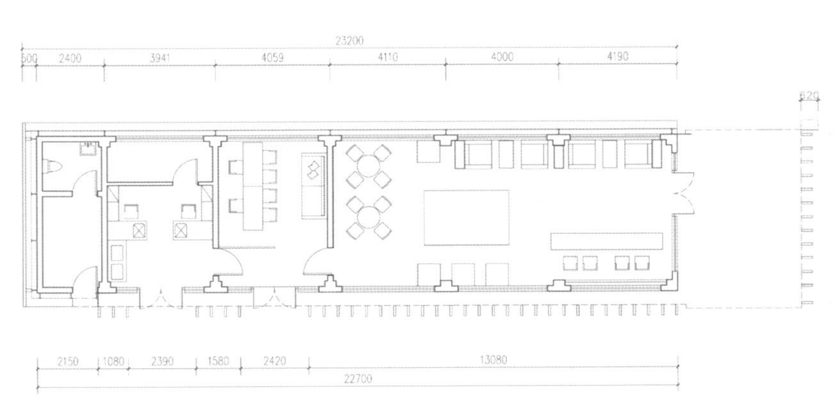
23200
500
2400
3941
4059
4110
4000
4190
620
2150
1080
2390
1580
2420
13080
22700

吉林长春东北师大附小益田幼儿园

吉林 长春 | 2010

Yitian Kindergarten Attached to Northeast Normal University, Changchun

Changchun, Jilin | 2010

性质：教育建筑
用地面积：35 071 m^2
建筑面积： 19 037 m^2
开发商：益田集团
主题词：开放式教学；自然温情

Property: Educational Architecture
Site Area: 35,071 m^2
Floor Area: 19,037 m^2
Client: Yitian Group
Keywords: Open Teaching; Natural and Warm

设计旨在创造自然温情、安全舒适的校园环境。校园空间丰富，以开放空间为核心布置功能空间，以此最大化缩短幼儿步行距离。同时扩大建筑与自然的接触面，创造能积极引导幼儿与自然交流及室外活动的过渡空间。建筑立面色彩跳跃，线条富于动感。

The project aims to create a natural and warm, comfortable and safe campus. In the rich campus space, the design arranges functional space with open space as the center to shorten children's walking distance. The design expresses the connection between the building and its natural surroundings, creating a healthful and open studying environment for children. The building's facade is finished with the bright, fresh, and dynamic colour and pattern.

海口龙华区丁村概念设计

海南 海口 | 2009

Concept Design of Ding Village in Longhua District, Haikou

Haikou, Hainan | 2009

性质：物流园
用地面积：836 488.62 m²
建筑面积：1 464 336.49 m²
开发商：海南现代科技集团有限公司
主题词：一站式；生态型；单元化

Property: Logistics Park
Site Area: 836,488.62 m²
Floor Area: 1,464,336.49 m²
Client: Hainan Modern Technology Group Co., Ltd.
Keywords: One-stop; Ecotype; Cell

项目以当代城市规划学和现代物流理念为指导，将丁村全力建设成为极具活力的都市节点，打造海南地区最具规模和影响力的居家主题购物中心，构建海口市极富文化内涵的个性城市景观。

With the guidance of contemporary urban planning and modern logistics concept, Ding Village is fully built into an extremely dynamic urban node, and the largest and the most influential home-theme shopping center in Hainan. The project constructs a unique urban landscape with rich cultural connotation in Haikou.

风翔路
龙昆南路
城南路
运大道
货运
龙昆南路
23F
27F

合肥华润置地澜溪镇

安徽 合肥 | 2009

CR Land French Annecy Town, Hefei

Hefei, Anhui | 2009

性质：住宅
用地面积：2 794.00 m²
建筑面积：9 902.55 m²
开发商：华润置地（合肥）有限公司
主题词：地景；退台；大平层

Property: Residence
Site Area: 2,974.00 m²
Floor Area: 9,902.55 m²
Client: CR Land Co.,Ltd.(Hefei)
Keywords: Landscape; Terrace; Flat Villa

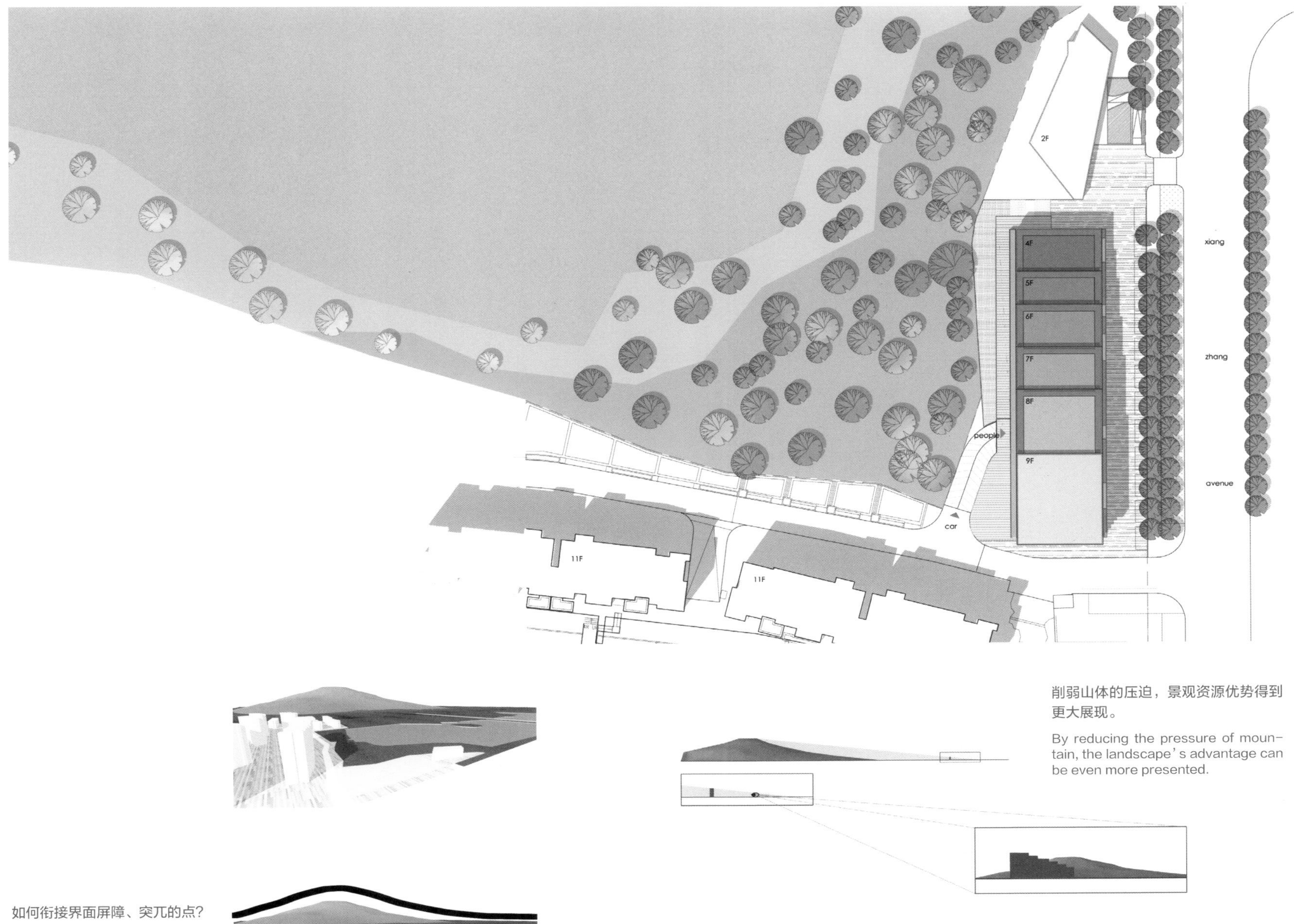

削弱山体的压迫，景观资源优势得到更大展现。

By reducing the pressure of mountain, the landscape's advantage can be even more presented.

如何衔接界面屏障、突兀的点？

How to link up the interface shielding and single commercial building?

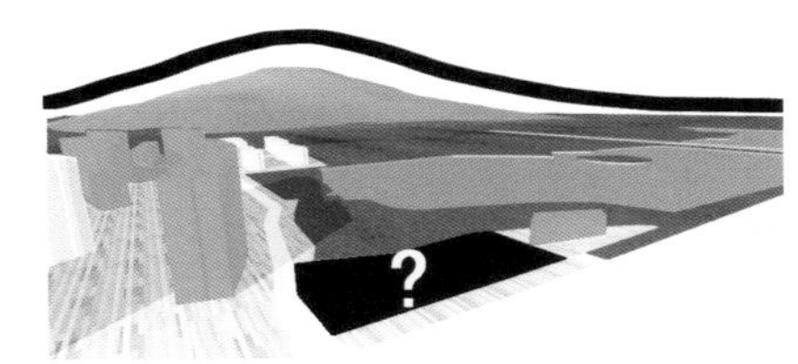

项目位于合肥高新开发区，周边交通便捷，自然景观资源丰富，当地政府以及建设方提出了建筑体量需与山、湖等自然环境相协调的设计要求。建筑整体构思采取退台的概念，与山体的自然曲线完美呼应，在创造自然流畅的建筑群衔接的同时，提供面向山水绝好风景的空中平台。

The project, located in Hefei Hi-tech Development Zone, has convenient transportation and abundant natural landscape resources. The local government and the client propose that the architecture should coordinate with the mountain and lake around. The concept of terrace perfectly responds to the natural curve of the mountain, creating a natural and smooth cohesion of building groups and providing a sky platform facing the beautiful scenes.

流畅承接已有建筑，与景观呼应。

It connects the existing buildings smoothly, corresponding to the landscape.

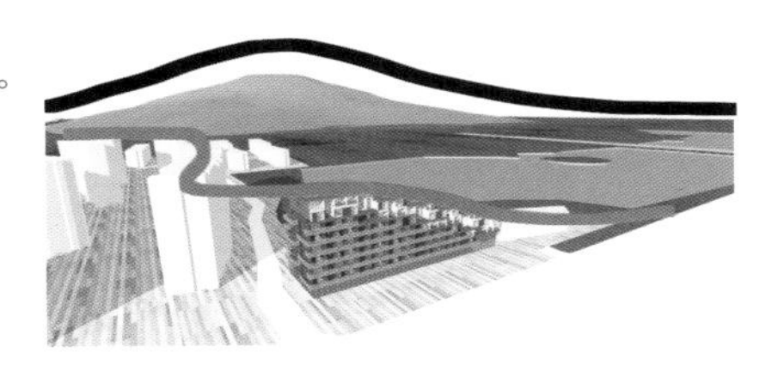

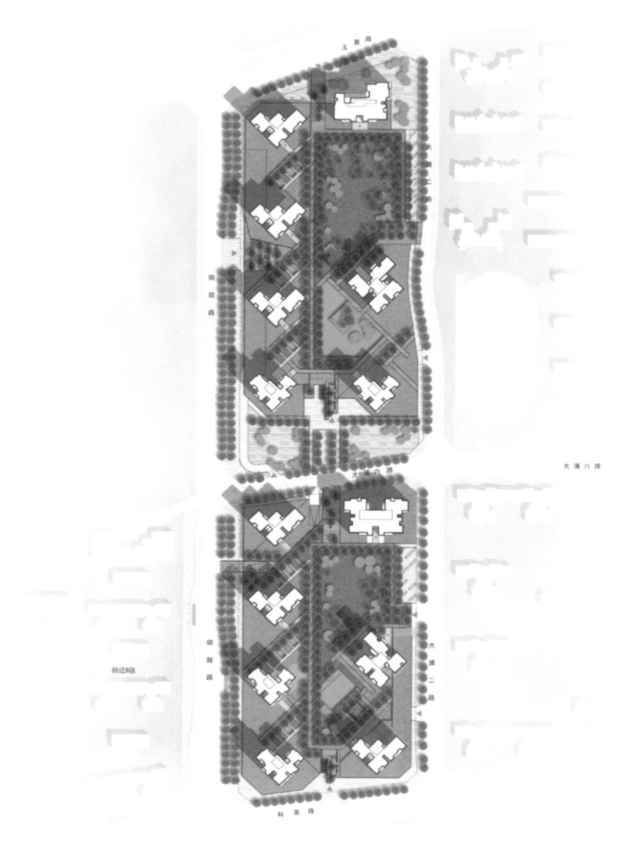

深圳华润城二、三期

广东 深圳 | 2013

Phase 2 & 3 of CR City, Shenzhen

Shenzhen, Guangdong | 2013

性质：住宅
用地面积：27 195 m^2
建筑面积：25 168 m^2
开发商：华润置地（深圳）有限公司
主题词：都市感；当代艺术风格；自然地景

Property: Residence
Site Area: 27,195 m^2
Floor Area: 25,168 m^2
Client: CR Land Co., Ltd. (Shenzhen)
Keywords: Urban Sense; Contemporary Artistic Style; Natural Landscape

各地块七栋住宅呈围合之势，绝大多数户型朝内向中心庭院，中心庭院贯通南北，形成U形的内部景观通廊。从规划整体来看，建筑形体简洁明快，与公建区形象相协调。远看城市界面，天际线高低起伏，融为一体，而自身不乏流露出当代艺术风格。里面通过深浅建筑体块，营造出完整而又具有动人韵律的现代都市建筑群落。

There are seven residential blocks which show the enclosing form. Most of the houses are facing inward toward the central courtyard which runs through the north and south, forming the U-shaped internal landscape corridor. In terms of the overall planning, the architectural form is simple and lively, which is in harmony with the image of the public construction area. From the perspective of the city interface, the skyline rises and falls harmoniously with contemporary artistic style. Through the deep and shallow building blocks, the whole modern urban building complex with moving rhythm is created.

深圳创茂侨香公馆

广东 深圳 | 2008

Qiaoxiang Mansion, Shenzhen

Shenzhen, Guangdong | 2008

性质：住宅
用地面积：27 467.36 m^2
建筑面积：224 946.00 m^2
开发商：广东创茂建设工程有限公司
主题词：高容积率；精致合理的户型设计；公建化立面

Property: Residence
Site Area: 27,467.36 m^2
Floor Area: 224,946.00 m^2
Client: Guangdong Chuangmao Construction Co., Ltd.
Keywords: High Plot Ratio; Delicate and Reasonable Unit Design; Facade of Public Building

项目充分利用基地的进深、面宽进行围合式布局，营造内部景观花园，实现高容积率要求。各期可通过架空层的空中花园联系，实现小区结构整体统一及系统延续。建筑布置结合户型单体设计，精致合理的户型设计最大限度地获得了良好朝向及景观。公建化的立面风格简洁明快，符合当代城市发展的风貌要求，同时增加社区的识别性。

To achieve high plot ratio, the project adopts an enclosing layout and creates an inner landscape garden by fully utilizing the depth and width of the site. Each phase is connected to another through an open air garden in order to maintain a holistic internal relationship. Buildings are arranged according to views and solar shading. The good orientation and view of the landscape are successfully achieved through delicate and reasonable unit design. The concise and commercial facade of public building accords with the requirements of modern urban development, and enhances the community' s identification.

侨

绵阳仙海湖住宅区

四川 绵阳 | 2010

Xianhai Lake Residential Area, Mianyang

Mianyang, Sichuan | 2010

性质：住宅
用地面积：209 024 m²
建筑面积：117 115 m²
开发商：四川长虹电子集团有限公司
主题词：山地别墅区；现代中式；自然亲和

Property: Residence
Site Area: 209,024 m²
Floor Area: 117,115 m²
Client: Sichuan Changhong Electronics Group Co.,Ltd.
Keywords: Villa Area in Mountains; Modern Chinese Style; Natural and Affiliative

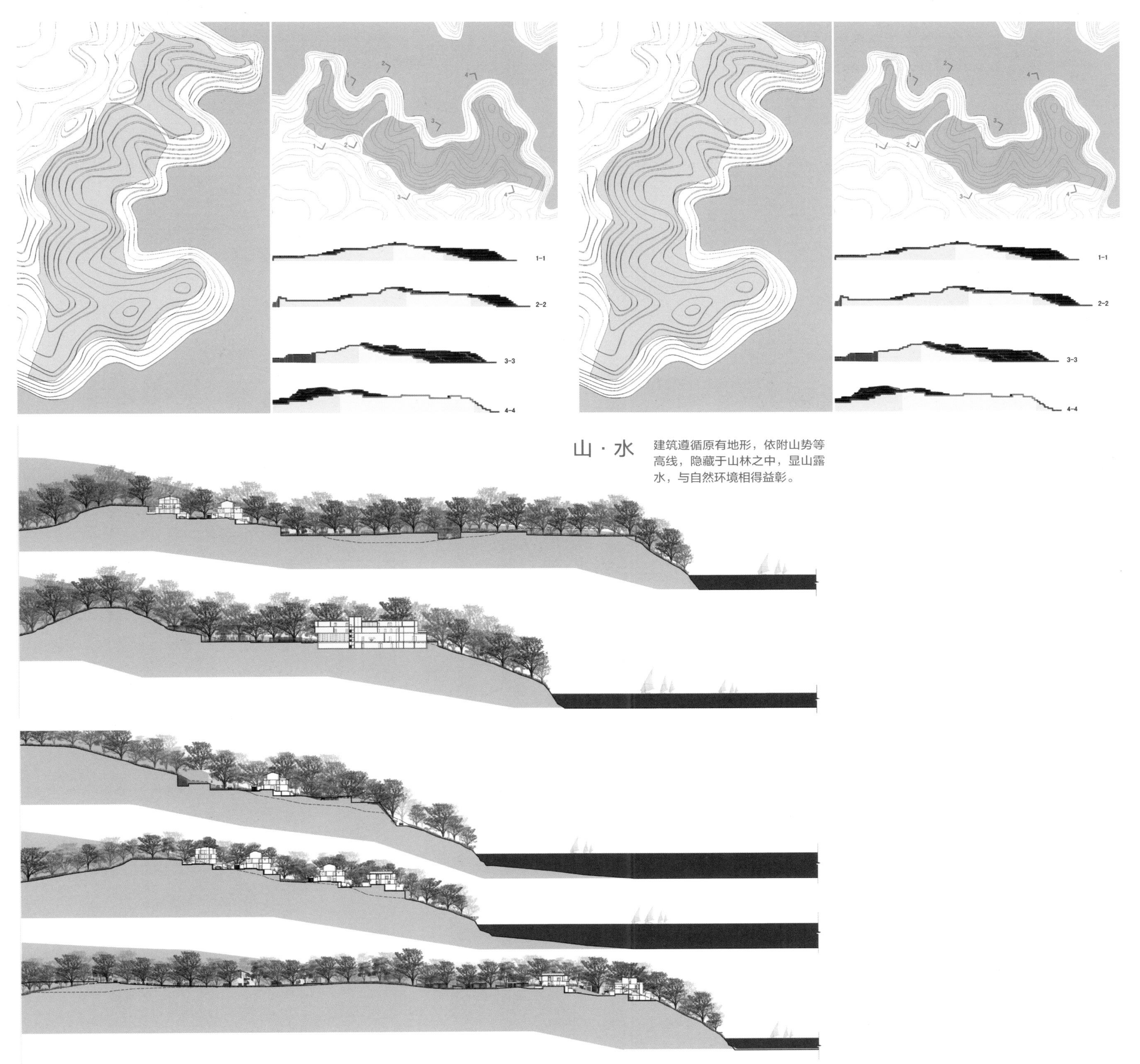

山 · 水

建筑遵循原有地形，依附山势等高线，隐藏于山林之中，显山露水，与自然环境相得益彰。

规划布局师法自然，顺应地形地貌，由南往北，构建由高至低、依山傍水的总体格局。建筑群体形成错落有致的天际轮廓，退台形态的建筑群落从视觉上使小区的尺度更加亲和。现代中式的设计手法使建筑立面简约、纯粹。建筑尺度宜人，隐身于自然的树林之中，与原生态自然景观相得益彰。

Inspired by nature, the planning of this project complies with the original topography to create the high-low overall layout from south to north, which is near the mountain and by the river. The skyline visually brings the buildings with terrace closer to the human. The facade design adopts modern Chinese style, which expresses concise and pure sense. The building hides easily within the natural forest with pleasant scale and is well matched with the natural surroundings.

海南三亚金茂海景花园

海南 三亚 | 2009

Jinmao Seascape Garden in Sanya, Hainan

Sanya, Hainan | 2009

性质：住宅
用地面积：11 364 m²
建筑面积：90 655 m²
开发商：广东创茂建设工程有限公司
主题词：高容积率；海景与建筑；精致合理的户型设计

Property: Residence
Site Area: 11,364 m²
Floor Area: 90,655 m²
Client: Guangdong Chuangmao Construction Co., Ltd.
Keywords: High Plot Ratio; Seascape and Building; Design of Exquisite and Proper Unit

建筑布局运用板点结合的方式，在东、北、西三面对称布置两栋板式高层住宅；南面结合楔形的地形特点，通过有针对性的户型平面设计，布置一栋点式高层住宅，满足高容积率要求。精致合理的户型设计，将海景与建筑充分融合。

立面设计突出节奏感，借助阳台引入波浪元素，赋予建筑形象浓厚的滨海特色。

This residential group is constructed using slab-type and point-type methods. It symmetrically allocates two panel-type high-rise residences in the east, north, and west. A single point-type tower, which contains designed housing units, is located on the southern side to interact with the site boundary. The design fulfills the client's request of high plot ratio by harnessing the last usable land of the site. The design of exquisite and proper unit fully integrates the seascape and buildings.

The facade design emphasizes the rhythm, and introduces wavy element to the balcony design, which makes the building full of strong coastal feature.

成都华润凤凰城

四川 成都 | 2007

CR Phoenix City, Chengdu

Chengdu, Sichuan | 2007

性质：住宅
用地面积：40 428 m²
建筑面积：225 191 m²
开发商：华润置地（成都）有限公司
主题词: 围合式规划布局；户型设计；亲和现代

Property: Residence
Site Area: 40,428 m²
Floor Area: 225,191 m²
Client: CR Land Co., Ltd. (Chengdu)
Keywords: Enclosing Layout; Unit Design; Affiliative and Contemporary

规划采用围合式布局，充分利用周边城市景观，营造大尺度社区景观庭院，住户均可获得良好的景观资源。十字点式高层住宅的布置及独特的景观设计，形成差异化的地块肌理及建筑形态，使社区具备识别性。人车分流的交通规划突显人文关怀。户型设计满足当地气候特点及市场要求，注重空间品质。立面设计采用亲和、适度、现代的建筑语言，营造独具特质的住区形象，建造高品质的居住小区。

The planning of the project adopts an enclosing layout and makes full use of the surrounding urban landscape to create large scale community landscape gardens. The layout of cross-sectional high-rise buildings and the unique landscape design create an environment that best meets the needs of the community. The separation of people and vehicles highlights the humanistic concern. The unit design adapts to the local climate and market demand, and focuses on the quality of space. The facade design uses an affiliative, moderate and contemporary architectural language to create a unique image of community and a high-quality residential district.

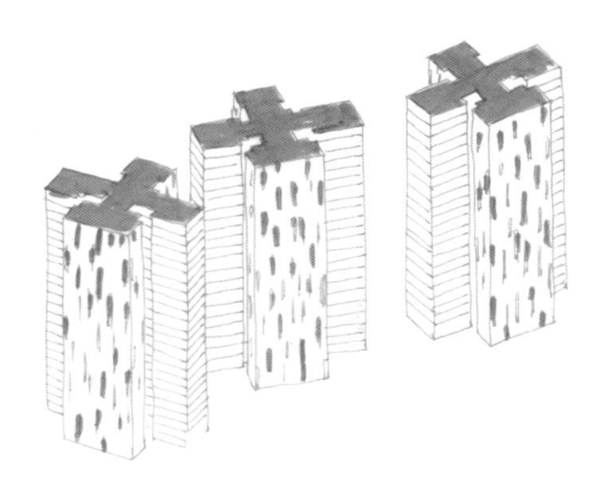

Facade conception for Phoenix City phase 2

凤凰城二期立面设计概念

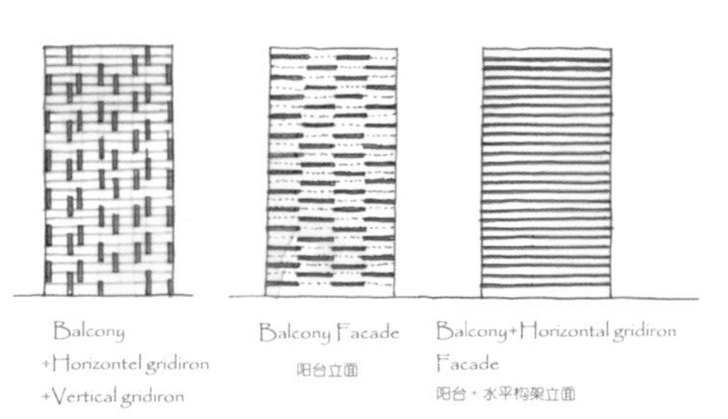

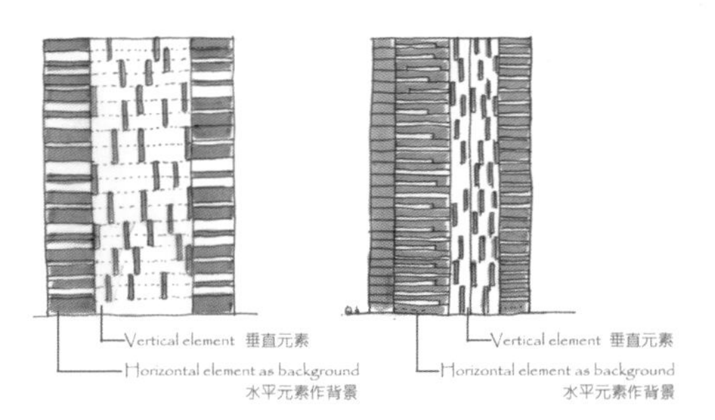

乐山嘉州新城

四川 乐山 | 2007

Jiazhou New City, Leshan

Leshan, Sichuan | 2007

性质：住宅
用地面积：61 965 m²
建筑面积：107 744 m²
开发商：乐山新业置地发展有限公司
主题词：总体设计；品质与成本；完成度

Property: Residence
Site Area: 61,965 m²
Floor Area: 107,744 m²
Client: Leshan Xinye Land Development Co., Ltd.
Keywords: Master Design; Quality and Cost; Degree of Completion

设计从总体规划角度考虑住区立面，通过系统化的建筑语言和色彩规划，改善了原建筑单元之间的关系。设计着重对宅间景观带山墙及檐口进行处理，使得单元间具备辨识性，避免小区空间同质化的同时，也使得原本无序的立面变得井然有序。此外，项目还针对当地施工工艺设计立面节点，采用低成本建造方式，获得了高品质观感。

The design considers architectural facade from the master planning, and improves the relationship between original buildings through systematic architectural language and colour planning. The design focuses on the gables, eaves and landscape between two residences, making the building units more distinctive. The design avoids homogenization of the community, and brings a level of order and harmony to the old disordered facades. The facade is designed by local construction techniques and uses a low-cost construction method to achieve high-quality sense.

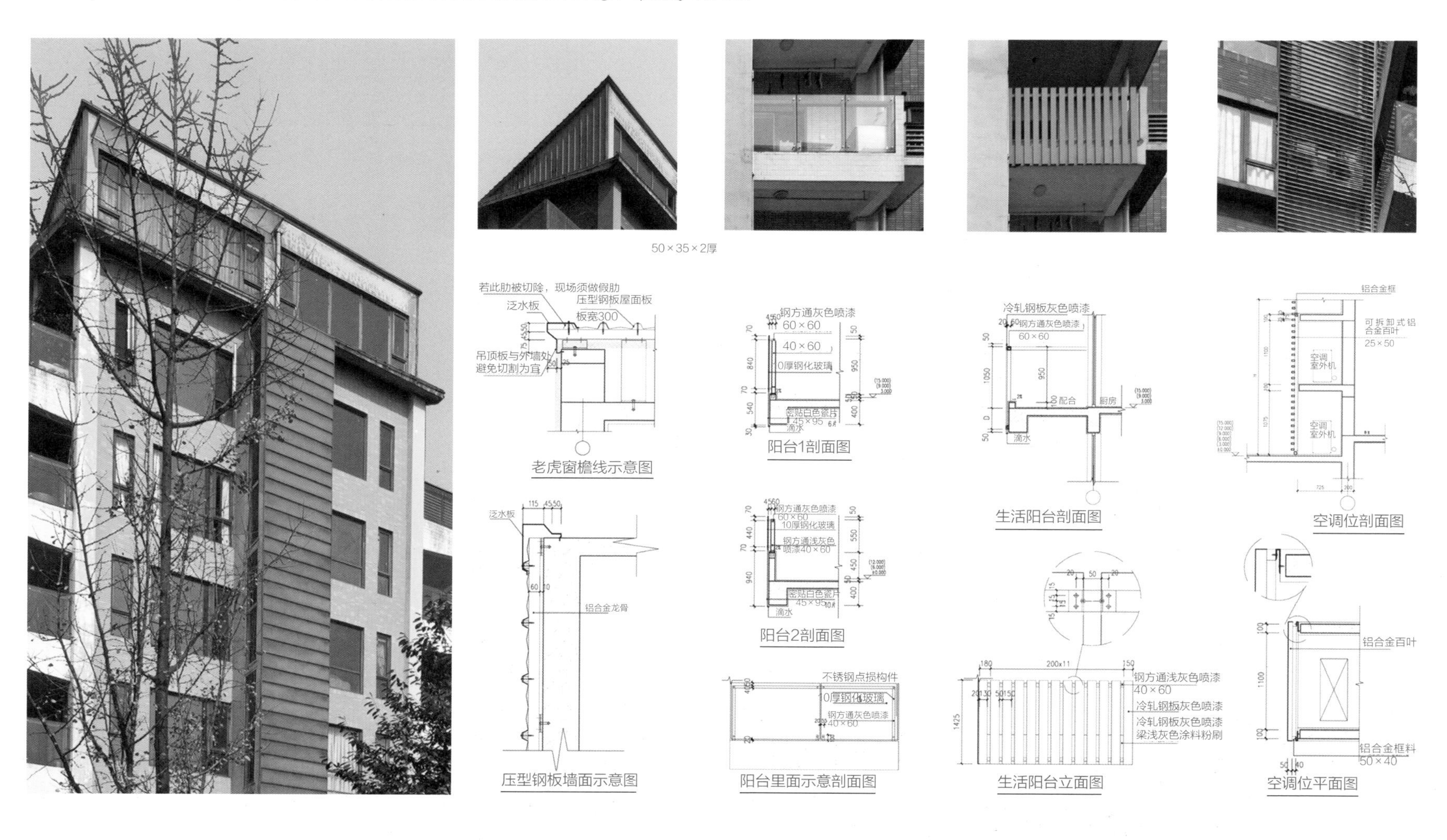

惠州中航屿海公寓

广东 惠州 | 2012

Zhonghang Yuhai Apartment, Huizhou

Huizhou,Guangdong | 2012

性质：公寓
用地面积：6 800 m²
建筑面积：7 736.52 m²
开发商：中航地产股份有限公司
主题词：依山就势；合理经济；尺度与细节

Property: Apartment
Site Area: 6,800 m²
Floor Area: 7,736.52 m²
Client: Zhonghang Land Co.,Ltd.
Keywords: Leaning against the Mountains; Reasonable and Economical; Scale and Detail

根据项目功能要求及周边环境特征，建筑依山而建，呈现出层层跌落、参差灵动的总体形态。

立面设计摒弃多余的装饰，运用现代简洁的设计手法，通过色彩划分及体块构成，形成既符合项目定位又与周边环境相协调的建筑表象。

According to the functional requirements of the project and the characteristics of the surrounding environment, the building leans against the mountains, showing the whole form of falling layers and agility.

Using modern concise design techniques and through color division and block composition, the facade design discards redundant decoration to form the building image which meets the requirements of the project and is coordinated with the surroundings.

深圳南澳下沙住宅区

广东 深圳 | 2011

Xiasha Residential Area in Nan’ao, Shenzhen

Shenzhen, Guangdong | 2011

性质：住宅
用地面积：132 000 m²
建筑面积：149 670 m²
开发商：广东创茂建设工程有限公司
主题词：滨海山地住宅；建筑与海景；现代简约；自然

Property: Residence
Site Area: 132,000 m²
Floor Area: 149,670 m²
Client: Guangdong Chuangmao Construction Co., Ltd.
Keywords: Coastal Residence in Mountains; Building and Seascape; Modern and Simple; Natural

该项目充分利用海景资源，并结合山地条件打造典型的滨海山地住宅。别墅分布在面朝大海的南向坡地，顺应山地地形呈退台布局，每户的主要房间均可观海。高层公寓沿北面山顶边界布置，舒展的走势和良好的朝向，使得户户朝海。高层公寓与别墅区在中间区域围合出开阔的山顶花园。建筑形式力求体现滨海特征及现代简约自然之美。

According to the local conditions of the mountain, the project constructs typical coastal residence in mountains by taking full advantage of the seascape. The cottages spread on the south slope facing the sea, and present the layout of terrace by following the topography of mountains, which ensures that the main rooms of each house are provided with the best view to the sea. The high-rise apartments are built along the northern border. Their layout effectively considers spacing and orientation with regards to the view to the ocean. A garden on the top of the mountain is enclosed by the apartments and villas. The architectural form strives to present the coastal features and modern, simple and natural beauty.

长沙五矿龙湾国际社区

湖南 长沙 | 2010

Longwan International Residence, Changsha

changsha, Hunan | 2010

性质：住宅
用地面积：266 145 m²
建筑面积：644 858 m²
开发商：五矿建设（湖南）旷代房地产开发有限公司
主题词：混合社区；法式风情；北美风格；哈佛主题小镇

Property: Residence
Site Area: 266,145 m²
Floor Area: 644,858 m²
Client: Minmetals Land Co.,Ltd.
Keywords: Mixed Community; French Style; North American Style; Harvard-Theme Town

项目在南北和东西向两条景观中轴线上组织点式及板式高层住宅。多层建筑在由南往北、从低至高的总体格局下，规划出清新的东西向带状组团，形成层次丰富、有秩序的住区空间。

一至三期高层立面采用法式风格，开创性地使用混凝土屋面板，以涂料饰面建造高层建筑坡屋顶，有效地控制了建设成本，打造魅力恒久的国际化高档住区。四期项目结合高端教育配套设施，在立面中植入来自北美的文化符号，打造哈佛主题小镇。

The project constructs the point-style and slab-style high-rise buildings on two north-south and east-west symmetric axes. In the overall pattern from low to high and from south to north, the multi-storey building plans a fresh east-west belt group to form a rich and orderly residential space.

The facades of Phase 1 to 3 adopt a French style, pioneering the use of painted concrete roof panels, and the sloping roofs of the high-rise building. The program effectively controls the construction costs while creating a charming and noble international residential area. Combined with the high-end educational facilities, the phase 4 implants cultural symbols from North America in the facade to create a Harvard-theme town.

长沙华润置地广场

湖南 长沙 | 2013

CR Land Square, Changsha

Changsha, Hunan | 2013

性质：商住综合体

用地面积：189 174 m^2

建筑面积：608 936 m^2

开发商：华润置地（湖南）有限公司

主题词：综合社区；街区商业；经济型写字楼

Property: Commercial & Residential Complex

Site Area: 189,174 m^2

Floor Area: 608,936 m^2

Client: CR Land Co., Ltd. (Hunan)

Keywords: Comprehensive Community; Block Business; Economical Office Building

华润·置地广场

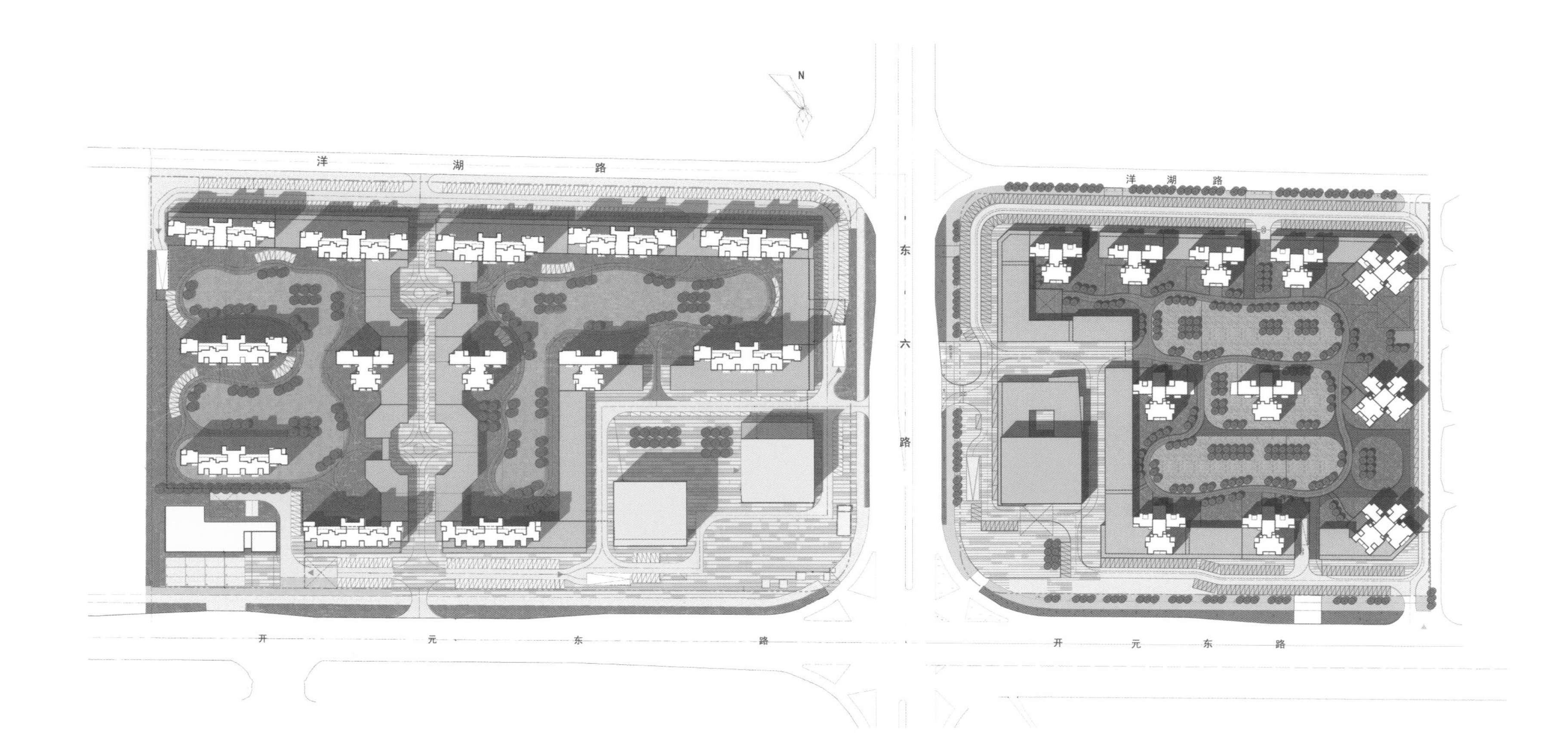

地块业态包括30%的写字楼及商业、70%的住宅。规划旨在塑造社区内部良好的空间环境，同时协调社区内外关系，实现住区与城市间和谐共生。在靠近地铁站的开元路及洋湖路街角布置高层写字楼，在营造城市公共空间的同时塑造社区地标形象。西地块分为东西两个组团布置，为地块创造了更多的临街界面，提高街铺的指标占比，以实现项目价值最大化的设计目标。

住宅户型设计严谨有序，尺度适宜。立面设计强调挺拔的竖向线条，写字楼、商业楼采用现代简约风格。

The types of the plot involve 30% office and commercial buildings, and 70% residential buildings in area. The planning aims to shape internal community space, and coordinate and implement a harmonious coexistence between settlements and the city. At the corner of Kaiyuan road and Yanghu road, close to the the subway station, high-rise office buildings are arranged to create a community landmark image while creating a public space of the city. The west block is divided into two east and west groups, which creates more street-facing interfaces for the plot, increases the proportion of street-shop indicators, and achieves the design goal of maximizing value.

The residential unit design is strict and adheres to local building codes and regulations. The facade design emphasizes the tall and straight vertical lines, and the office and commercial buildings adopt the contemporary simple style.

惠州五矿哈斯塔特山地住宅

广东 惠州 | 2014

Minmetals Hallstatt Hillside Residence, Huizhou

Huizhou, Guangdong | 2014

性质：住宅

用地面积：31 589.00 m²

建筑面积：16 304.19 m²

开发商：五矿建设(湖南)旷代房地产开发有限公司

主题词：极限的山地条件；创新小户型公寓；奥地利风情小镇

Property: Residence

Site Area: 31,589.00 m²

Floor Area: 16,304.19 m²

Client: Minmetals Land Co., Ltd.

Keywords: Extreme Mountain Condition; Innovative Small Apartment; Austrian-style Town

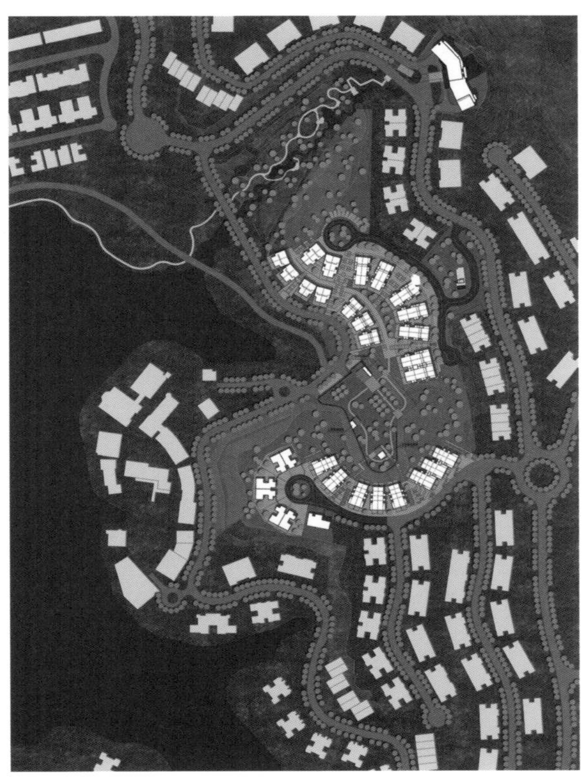

项目以奥地利风情小镇哈斯塔特为设计蓝本，在最陡坡度接近45度的山地条件下，因地制宜，以最少土方量化解巨大场地高差。充分利用湖景及场地关系，结合地形高差设计小户型公寓，层层跌落，每户均有面向湖景的露台花园。

The project draws inspiration from a Austrian-style town named Hallstatt. The site involves a 45 degree slope at most. The program aims to create a design that least alters the existing topography of the site. Combined with the topographical differences, the small-sized apartments are designed by mean of relationship between the lake and site, and each unit has a terrace garden facing the lake.

惠州五矿哈斯塔特度假酒店

广东 惠州 | 2012

Minmetals Hallstatt Resort Hotel, Huizhou

Huizhou, Guangdong | 2012

性质：酒店
用地面积：6 600 m²
建筑面积：11 443 m²
开发商：五矿建设(湖南)旷代房地产开发有限公司
主题词：在限制中设计；奥地利风情

Property: Hotel
Site Area: 6,600 m²
Floor Area: 11,443 m²
Client: Minmetals Land Co., Ltd.(Hunan)
Keywords: Design in Constraints; Austrian Style

建筑延续旁边小镇的奥地利风情，复原城堡的意象，为客人提供一种独特的居住体验。

通过对建筑体量的分解划分，创造出轻盈漂浮的感觉，结合水景营造出一种梦幻的意境。

建筑上部采用城堡意象，并破除传统城堡的封闭感，提高了居住的舒适感和景观观赏度，利用屋顶空间提供特别的居住感受。

The building continues to adopt the Austrian style of the town next to it, restoring the castle's image and providing guests with a unique living experience.

Through the division of the building volume, the project creates a light floating feeling, which is combined with the water scene to create a dreamy mood.

The upper part of the building adopts the castle's image and breaks the enclosure of traditional castle to improve the comfort of living and the effect of the landscape, and uses the roof space to provide a special feeling of living.

惠州五矿哈斯塔特矿山公寓

广东 惠州 | 2014

Minmetals Hallstatt Mine Apartment, Huizhou

Huizhou, Guangdong | 2014

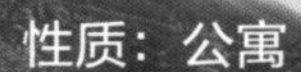

性质：公寓
用地面积：3 917.25 m^2
建筑面积：7 233.22 m^2
开发商: 五矿建设(湖南)旷代房地产开发有限公司
主题词：景观山居；建筑与自然

Property: Apartment
Site Area: 3,917.25 m^2
Floor Area: 7,233.22 m^2
Client: Minmetals Land Co., Ltd.(Hunan)
Keywords: Landscape Mountain; Architecture and Nature

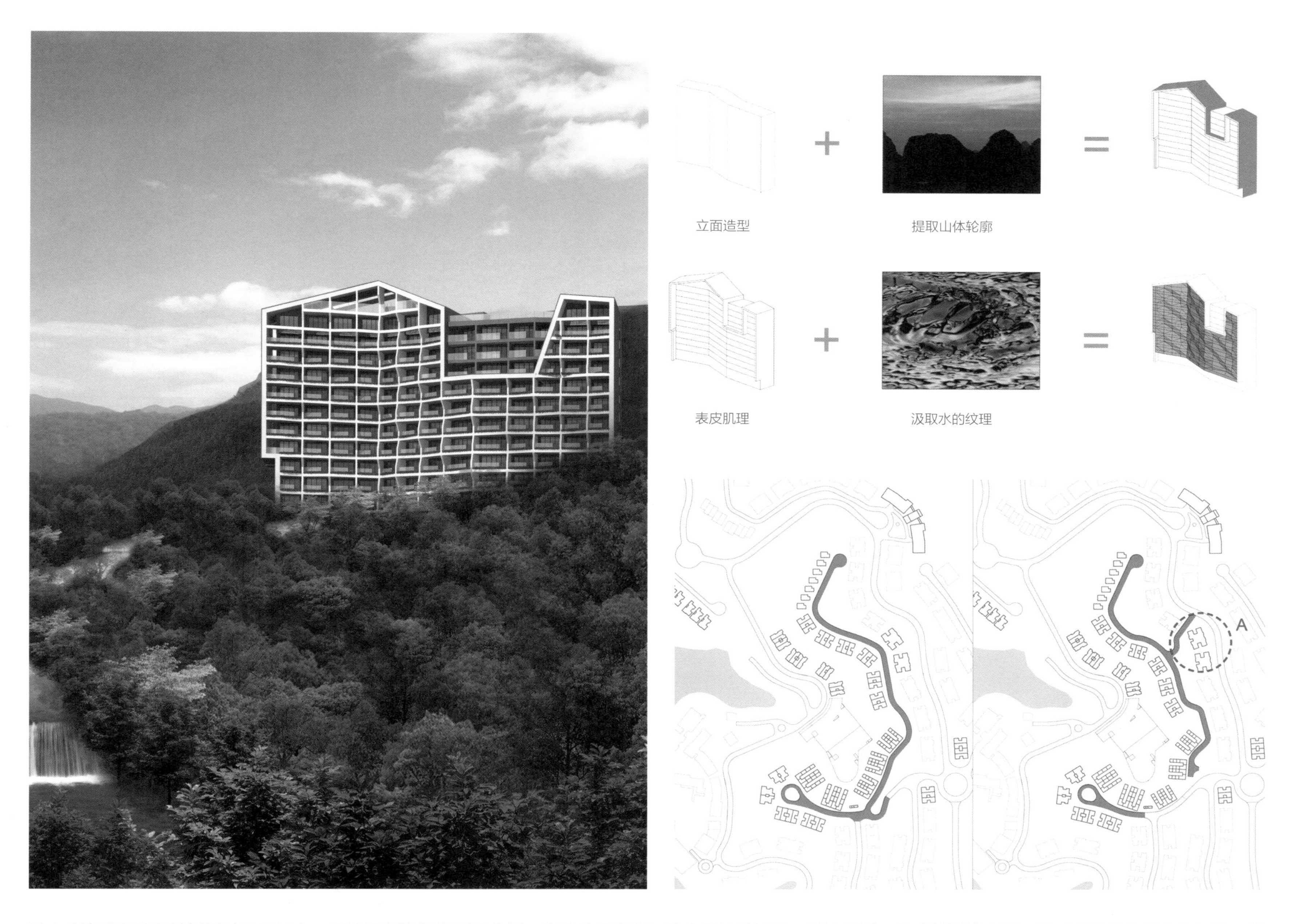

矿山公寓项目位于哈斯塔特风景区内，用地进深较短且形状曲折。为了使公寓更加符合用地的特征，设计采用三段式的折线布局，并对项目用地所临的场地和道路的关系进行梳理，重新调整了道路的形态。因为项目依山而建，所以建筑的轮廓线模拟山体曲折蜿蜒的形态。又因为建筑面对湖景，在立面形式上，采用了水的肌理，利用阳台有规律的及模数的变化表达水的意向。整个建筑形式上既传递了山的阳刚，又诠释了水的柔美，将成为景区内一处极具特色的标志建筑。

The project of mine apartment is located in the Hallstatt scenic spot, and the land's depth is short and the shape is tortuous. In order to make the apartment more consistent with the characteristics of the land, the design adopts a three-stage folding line layout, and readjusts the shape of the road after combing the relationship between the site and road. Because the project is built along the mountains, the outline of the building simulates the characteristics of the twists and turns of the mountains. And because the building faces the lake, the water's texture is adopted in the form of facades, and regular and modular changes are used on the balcony to express the intention of water. The entire architecture not only conveys the masculinity of the mountains, but also interprets the softness of the water, which become a distinctive symbolic building in the scenic spot.

合肥幸福联盟紫郡

安徽 合肥 | 2009

Happiness Union Purple County, Hefei

Hefei, Anhui | 2009

性质：住宅
用地面积：30 842 m²
建筑面积：77 448 m²
开发商：幸福联盟投资有限公司
主题词：北欧立面风格；建筑与环境的融合

Property: Residence
Site Area: 30,842 m²
Floor Area: 77,448 m²
Client: Union Of Happiness Investment Co., Ltd.
Keywords: Nordic Facade; Integration of Building and Environment

规划布局结合场地形态特征，顺应项目定位需求，由南往北呈现从低到高的建筑格局。建筑群落形态蜿蜒，天际轮廓错落有层次，充分实现建筑与环境的融合。退台设计使建筑获得最大的开阔视野，也使建筑的尺度更加亲和。

建筑立面呈现北欧风格，高贵典雅且不失鲜明。

The planning considers the topographic feature and the project's positioning, presenting a low-high architectural arrangement from the north to the south. The winding building group and its skyline with layers fully integrate the building and environment. The design of terrace gives the building maximum view while making the building scale more affiliative.

The building's facade adopts the Nordic style, which is noble, elegant and distinctive.

湘潭金侨夏威夷公馆

湖南 湘潭 | 2017

Jinqiao Hawaii Mansion, Xiangtan

Xiangtan, Hunan | 2017

性质：住宅
用地面积：161 422 m²
建筑面积：429 079 m²
开发商：湘潭滨江置业有限公司
主题词：景观利用最大化；利用地形规划布局；法式风格

Property: Residence
Site Area: 161,422 m²
Floor Area: 429,079 m²
Client: Xiangtan Binjiang Property Co., Ltd.
Keywords: Maximum Use of Landscape; Planning and Layout by Topography; French Style

JINQIAO2018

项目位于湘潭市九华经济开发区，南侧紧邻湘江。充分发挥地形优势，使得江景利用最大化是规划设计的核心理念。总设计采用“一轴一带三组团”的规划手法组织小高层住宅及高层住宅。住宅立面采用法式古典与现代简约相结合的设计手法，大气、精致、典雅。

The project is located in Jiuhua Economic Development Zone of Xiangtan City, and its south side is adjacent to Xiang River. Fully taking advantage of the topography and maximizing the utilization of river are the core concept of the planning and design. In the general design, small-scale high-rise residential buildings and high-rise residential buildings are organized by the planning method of “one axis, one belt, three groups”. The facade design of the house is a combination of French classicism and modern minimalism, which is magnificent, exquisite and elegant.

图书在版编目（C I P）数据

华域普风建筑作品选 / 华域普风著 . —天津：天津大学出版社，2018.10
ISBN 978-7-5618-6274-2

Ⅰ . ①华… Ⅱ . ①华… Ⅲ . ①建筑设计—作品集—中国—现代 Ⅳ . ① TU206

中国版本图书馆 CIP 数据核字 (2018) 第 241713 号

华域普风建筑作品选
Huayupufeng Jianzhu Zuopinxuan

图书策划　杨云婧
文字编辑　李　轲
美术设计　高婧祎
图文制作　天津天大乙未文化传播有限公司
编辑邮箱　yiweiculture@126.com
编辑热线　188-1256-3303

出版发行　天津大学出版社
地　　址　天津市卫津路 92 号天津大学内（邮编：300072）
电　　话　发行部 022-27403647
网　　址　publish.tju.edu.cn
印　　刷　廊坊市瑞德印刷有限公司
经　　销　全国各地新华书店
开　　本　250mm × 255mm
印　　张　13 2/3
字　　数　129 千
版　　次　2018 年 10 月第 1 版
印　　次　2018 年 10 月第 1 次
定　　价　198.00 元